Manufacturing Technologies for Machines of the Future

Springer-Verlag
Berlin Heidelberg
GmbH

Anatoli Dashchenko (Editor)

Manufacturing Technologies for Machines of the Future

21st Century Technologies

With 431 figures and 41 tables

Springer

Professor Dr. Sc. ANATOLI I. DASHCHENKO

Moscow State Technical University MAMI
B. Semenovskaya St. 38
105839 Moscow

Additional material to this book can be downloaded from http://extras.springer.com.
ISBN 978-3-642-62822-1 ISBN 978-3-642-55776-7 (eBook)
DOI 10.1007/978-3-642-55776-7

Cataloging-in-Publication Data applied for
A catalog record for this book is available from the Library of Congress.
Bibliographic information published by Die Deutsche Bibliothek
Die Deutsche Bibliothek lists this publication in the Deutsche Nationalbibliografie;
detailed bibliographic data is available in the Internet at http://dnb.ddb.de

Originally published by Springer-Verlag Berlin Heidelberg New York in 2003
Softcover reprint of the hardcover 1st edition 2003

http://www.springer.de

Coverdesign: deblik, Berlin
62/3020 UW Printed on acid-free paper - 5 4 3 2 1 0 -

List of Authors

Anderl, R., Prof., Dr. Eng.
Technical University Darmstadt
Petersenstr. 30, 64287 Darmstadt

Dalsci, A.M., Prof., Dr. Sc.
Bauman Moscow State Technical University
2nd Baumanskaya St. 5, 107005 Moscow

Dashchenko, A.I., Prof., Dr. Sc.
Moscow State Technical University MAMI
B. Semenovskaya St. 38, 105839 Moscow

Dashchenko, O.A., Dipl., Eng. Dr. techn.
DaimlerChrysler AG
Epplestr. 225, 70546 Stuttgart

Emelyanov, V.V., Prof., Dr. Sc.
Bauman Moscow State Technical University
2nd Baumanskaya St. 5, 107005 Moscow

Feldmann, K., Prof., Dr. Eng.
Friedrich-Alexander-University
Egerlandstr. 7-9, 91058 Erlangen

Frolov, K.V., Prof., Dr. Sc.
Academy of the Russian Academy of Sciences
Mechanical Engineering Research Institute
M. Haritonievski St. 4, 101990 Moscow

Gausemeier, J., Prof., Dr. Eng.
University of Paderborn
Fuerstenallee 11, 33102 Paderborn

Gornev, V.F., Prof., Dr. Sc.
Bauman Moscow State Technical University
2nd Baumanskaya st. 5, 107005 Moscow

Heisel, U., Prof., Dr. Eng.
University of Stuttgart
Holzgartenstr. 17, 70174 Stuttgart

Hatamura, Y., Prof., Dr. Eng.
University of Tokyo
7-3-1 Hongo, Bunkyo-ky, 113-8656 Tokyo

Hügel, H., Prof., Dr. Eng. habil.
University of Stuttgart
Pfaffenwaldring 43, 70569 Stuttgart

Iassinovski, S.I., Prof., Dr. Sc.
Bauman Moscow State Technical University
2nd Baumanskaya St. 5, 107005 Moscow

Inaba, H., Dipl. Eng.
FANUC Ltd
Oshino-mura, 401-0597 Yamanashi

Jovane, F., Prof.
Politecnico di Milano
Direttore ITIA-CNR
Viale Lombardia 20/A, 20131 Milano

Kluev, N.N.
Corresponding Member of the Russian Academy of Sciences
MNPO "Spectr"
Usachova St. 35, 119048 Moscow

Koren, Y., Prof., Dr. Sc.
University of Michigan
2270 G.G. Brown Building, 2350 Hayward St.
Ann Arbor, Michigan 148109-2125

Mandelli, M., Dipl. Eng.
UCIMU-sistemi per produrre
Viale Fulvio Testi 128
2000/L461, 20092 Cinisello Balsamo, Milano

Mitsuishi, M., Prof., Dr. Eng.
University of Tokyo
7-3-1 Hongo, Bunkyo-ky, 113-8656 Tokyo

Moriwaki T., Doctor of Engineering, Prof.
Kobe University
1-1 Rokkodai-cho, Nada-ku, Kobe 657-8501

Nachapetjan, E.G., Prof., Dr. Sc.
Mechanical Engineering Research Institute
M. Haritonievski St. 4, 101990 Moscow

Nagao, T., Prof., Dr. Eng.
Kochi University of Technology
Tosayamada, 782-8502 Kochi

Nee, A.Y.C., Prof., PhD
National University of Singapore
10 Kent Ridge Crescent, 119260 Singapore

Nakao, M., Prof., Dr. Eng.
University of Tokyo
7–3–1, Hongo, Bunkyo-ky, 113–8656 Tokyo

Ong, S.K., Ass. Prof., PhD
National University of Singapore
10 Kent Ridge Crescent, 119260 Singapore

Panfilov, Y.I., Prof., Dr. Sc.
Bauman Moscow State Technical University
2nd Baumanskaya st. 5, 107005 Moscow

Peklenik, J., Prof., Dr. Eng. habil
University of Ljubljana
Askerceva 6, 61000 Ljubljana

Pollmann, W., Prof.
DaimlerChrysler AG
Production Technologies Research & Development
Epplestr. 225, 70546 Stuttgart

Pritschow, G., Prof., Dr. Eng.
University of Stuttgart
Seidenstr. 36, 70174 Stuttgart

Rehsteiner, F.H., Prof. Dr. Sc.
Institute for Machine Tools and Manufacturing
Swiss Federal Institute of Technology (ETH)
Tannenstr. 3, 8092 Zurich

Santochi, M., Prof., Dr. Eng.
University of Pisa
Via Bonanno Pisano, 25, 56126 Pisa

Schraft, R.D., Prof., Dr. Eng.
Fraunhofer Gesellschaft, IPA
Nobelstr. 12, 70569 Stuttgart

Schulz, H., Prof., Dr. Eng.
Technical University Darmstadt
Petersenstr. 30, 64287 Darmstadt

Sherbakov, V.V. †, Dr. Sc.
Mechanical Engineering Research Institute
M. Haritonievski st. 4, 101990 Moscow

Ulsoy, G.,
William Clay Ford Professor of Manufacturing
University of Michigan
2236 G. G. Brown Building, 2350 Hayward St.
Ann Arbor, Michigan 48109–2125

Van Brussel, H., Prof., Dr. Eng.
Katholieke Universiteit Leuven
Celestijnenlaan 300 B, 3001 Heverlee

Vasiliev, A.S., Prof., Dr. Sc.
Bauman Moscow State Technical University
2nd Baumanskaya st. 5, 107005 Moscow

Volchkevich, L.I., Prof., Dr. Sc.
Bauman Moscow State Technical University
2nd Baumanskaya st. 5, 107005 Moscow

Warnecke, H.J., Prof., Dr. Eng.
Fraunhofer Gesellschaft, President
Leonrodstr. 54, 80636 Munich

Westkämper, E., Prof., Dr. Eng.
Fraunhofer Gesellschaft, IPA
Nobelstr. 12, 70569 Stuttgart

Foreword

This is a unique edition in a new format: a combination of text and video material in a book and accompanying CDs. The idea to produce a work to represent the bottom line of scientific and industrial development at the beginning of the 21st century came from leading scientists.

This book undertakes analysis, description, and development prognoses of machine-building technologies and production automation (including examples of basic research of the development and optimization of real-world production processes, as well as methods and results of experimental research and presentation of some of the most modern and innovative manufacturing processes) which should form a basis for the automated production technologies of the future – in the 21st century.

This book offers an interdisciplinary presentation of unique material and combines for the first time theoretical and practical results of the last decades from the most important branches of machine building in industrial-developed countries (automotive industry, agricultural machine building, electrical engineering and electronics, machine-tools industry, aircraft industry, instrumental industry, control systems, and consumer goods industry) in a scientific-technical edition. This should make this book interesting to a wide range of readers. It is aimed at those who because of their knowledge and talent will become the elite engineers of their respective countries: doctoral and undergraduate students, to young prospective and qualified engineers, to advanced beginners and well-known scientists and researchers.

It is common to regard many engineers and scientists are the founders of the scientific–technical progress, but this is not so. Truly revolutionary breakthroughs are a prerogative of the technical elite, which represent less than 10% of specialists. Membership of this elite club is determined not only by natural talent but also by hard work and devotion to this field. These very categories of engineers developed the space and atomic industries, and modern electronics including computer techniques.

The quicker that scientific–technical progress is, the more complicated and diverse are the processes of creation and implementation of the new. The most progressive technical innovations come from a long and arduous creation pathway, and further maturation is accompanied by phases of unbelief and disappointments.

The authors hope that while reading this book – either all of it or only those parts of it which attract your attention – you will obtain not only a concrete knowledge about the most advanced achievements in the field of new technologies, but also a feel for the importance of the advanced technologies that should form the basis for automated production in the 21st century.

Reading this book should enable you to not only evaluate objectively the directions of development of the newest techniques and technologies, many of which have had and will have a great influence on the development of humanity, but to deepen your knowledge in those spheres where you will contribute to the technical progress of your own branches of human knowledge.

Of course, our book cannot cover all the problems associated with machine building, because these machines are so various in their implementation, structure, and even external appearance. They are being modernized so quickly such that new types of machines are appearing continuously so that even a huge encyclopedia would not be able to cover all the topics. In any case, encyclopedias are being published slowly, so that many new tendencies, constructions, and inventions appear in the intervening time. The same argument applies to advanced manufacturing technologies.

However, the common aspects of machine production, methods and technologies, problems and ways of solution in the field of production automation, management, quality control, usage of computer (information) technologies, virtual and augmented reality, and the influences of production technologies on the environment – all of which make up the contents of our book – are changing more slowly than progress in the actual construction of machines. That is why this book could be useful not only for creation of modern techniques but also for the development of examples of its future generations.

Good luck, dear reader!

Prof. W. Pollmann
Chairman of the Editorial Board
Vice President
Research, Materials, and Production
DaimlerChrysler AG

Preface

With this "book" – a term like "media complex" might be more appropriate – the authors have combined two modern ideas to give the reader an up-to-date view of the world of manufacturing. Its basic quality is derived from the choice of its contents and the preparation of the individual chapters by an international group of leading experts. In addition, it has been given a uniquely modern appearance in that the basic "book" is supplemented by CDs containing videos, graphics, and web pages to give the reader a yet more lively understanding of its contents.

Care has been taken to treat the three fundamental disciplines of manufacturing equally: the processes, the equipment, and all the peripheral items used to combine components into systems; these encompass all computer applications, including communication, networking and graphics.

The book is aimed at different people that are interested in manufacturing. By its broad coverage, it should appeal to the highly skilled scientific expert as much as to the experienced design engineer, and to students in their early years as well as to doctoral students. The book combines basic, relatively abstract concepts with practical examples and application reports.

Another idea pursued in this book represents, to some extent, a visionary view. At the onset of the 21st century (or, indeed, of the third millennium), we realize that manufacturing – seemingly mature and hampered by centuries of tradition – explodes with new concepts and ideas. Particularly fascinating is the fact that very old concepts, such as manufacturing an object within minutes of its design being finalized at a completely different place, are now realistic by combining disciplines such as CAD (computer-aided design), rapid prototyping, and metrology. Thus, learning about manufacturing involves learning about its potentials.

To help the reader find his way through the text, it has been grouped into five parts. Part I deals with a view on global manufacturing, thus emphasizing the widely varying external conditions for and expectations from manufacturing. In Part II some highly advanced processes (e.g., high-speed machining) and their combination to tightly woven systems of processes are discussed together with their virtual modeling. Part III shows examples and applications of this. After these discussions of the processes, Part IV is devoted to the equipment used to implement them where, again, real machinery (e.g., parallel kinematic machines) is discussed as well as global concepts like reconfigurable machines. The book closes with an outlook to future trends in Part V.

Happy reading!

Prof. Dr. Sc. A. Dashchenko
Scientific Editor
Head of Department for Complex Automation in Machine Building.,
Moscow State Technical University "MAMI", Russia

Contents

Part V: Future Trends

Part IV
Advanced Manufacturing Equipment

Chapter 18
New Machine Tools and Systems

M. Mandelli, T.Nagao, Y. Hatamura, M. Mitsuishi, M. Nakao

18.1
New Machine Tools for New Manufacturing Processes

M. Mandelli

18.1.1
Introduction

The machine tool industry is enjoying a period fast innovation, driven by end-users that have rediscovered the importance of production, after a long period during which finance and marketing seemed to be the sole instruments to reach industrial success. Innovaticn is also pushed by internal pressure: European manufacturers must keep proposing new solutions to face the evolution of a market that is increasingly competitive.

It is estimated that in the last ten years the number of functions and the quality content required for investment goods increased by 50%, while the market price available corresponding to the same product decreased by 30%, under the pressure held by the growing international competition and by market pressure.

In the case of machine tools, the ma2ket is particularly pressing. Machine tools are used to produce other machines; they are one basic factor of competitiveness for mechanical firms.

The machinery sector is characterized by the necessity of continuous research for improving competition in terms of product efficiency, flexibility, and velocity of response to the customers, particularly in western countries. Inside this context production means play a fundamental role.

In the case of large industrial groups, the needs connected to globalization are extremely important: the production sites are located in different countries, and each one must guarantee a level of efficiency suitable to its own survival inside the group; they must guarantee a common qualitative standard, capacity of cost reduction, and a level of flexibility good enough to respond to the changing of market needs.

Mainly these types of firms require "turn-key systems". It is always more and more difficult to find inside-the-company organization; people and skills able to define the fittings for the different production systems that, for the severe request of standard-series products with a high production rate, high productivity, and high efficiency, are too difficult to handle. Such an organization, able to manage production plants in a completely autonomous way, may be difficult to justify in term of costs.

18.1.2 Which Are the Customers' Needs?

18.1.2.1 Reliability

From a market study, recently realized in Great Britain with a large number of companies – users of machine tools, emerged very interesting indications regarding the arguments judged of major importance during the buying process of a new production machine. Objects of this particular study were milling machines and manufacturing centers, but the results obtained can be easily extended to different kinds of machines.

From the diagram of Fig. 18.1, it is clear that all the items connected with reliability are critical.

This is one of the consequences of "lean" organizations, where cost efficiency is obtained, also avoiding redundancy of production means, but in case of failure of one production machine, the cost to be considered is not only the cost of machine time, but the cost of the lack of production.

18.1.2.2 Technical Skill of the Suppliers

If we analyze the buying process of big end users, for example tier-one suppliers for the automotive industry, it clearly appears that inside the relation between users and manufacturers of machine tools, the capacity of the suppliers of developing the complete definition of the production plant, consider its application, is of capital importance: the major concern must be the object, the part, that the plant will produce. The skills needed to efficiently solve the problem of the end user have a paramount importance.

Strictly linked to this aspect is the capacity to supply "turn-key systems," and for the machine tool manufacturers, the main objects of the supply are not only intrinsic quality of the plant or of the machine, but more importantly the quality of the parts produced with the machine and the characteristics of the production cycle (particularly time and cost).

The capability to build up a production system in a very short time is another aspect of great importance today. This need is a consequence of the reduction of time to market persued by the most important manufacturers.

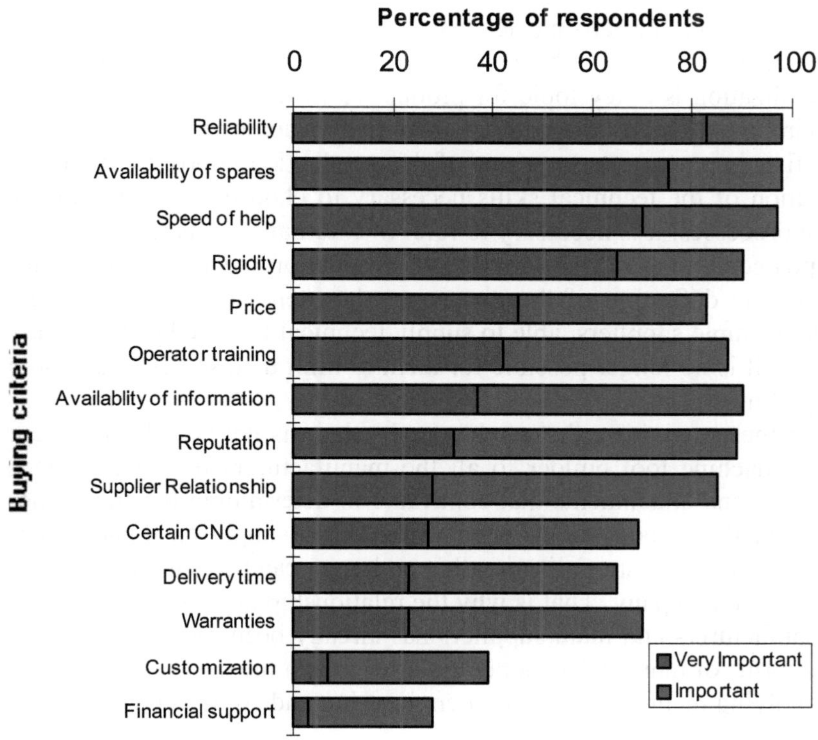

Fig. 18.1 Buying criteria for milling machines and machining centers.

To properly respond to this request, one possibility of reducing delivery time for "turn-key systems" is to increase the utilization of standard modules, easy to assemble inside customized systems.

The last aspect that must be considered by a machine tool manufacturer who wants to reach real customer satisfaction is service, which must be provided during the entire life cycle of the production plant. Timeliness and efficiency of interventions are of vital importance for the end user.

18.1.3
From Product to Process

We have seen that today the major concern of end users has shifted from the stand-alone machine to a more global vision that embraces the entire production process, made necessary to obtain a defined product with the desired quantitative and qualitative characteristics.

The users' attention is focused on the final result of machining; how this result is obtained is less important.

Companies are specializing, focusing their interests on their own products, which eventually could be included in a further more complex product.

Specialization is a key topic for productive firms, and in this competitive environment, being "the best" in one's field becomes a key of success.

The first important consequence of this production approach is the externalization of the technical skills necessary to choose and continuously improve processes: it is necessary to refer to external aspects to acquire the most updated competencies. Suppliers of production means are deeply involved in the definition of the processes; they tend to become partners more than simple suppliers, able to supply technical knowledge to the customer site. It is no longer possible for them to limit themselves to a simple delivery of machines.

Inside these activities, the whole supply chain is necessarily involved: from the machine tool builder to all the manufacturers of those components, tools, and attachments that contribute in determining the final function. Again, the amount of knowledge needed to supply an updated machine requires a great specialization that no longer can be found completely inside a single company. That is why the relationship between the machine tool manufacturers and main suppliers has already been modified, moving towards forms of more strict cooperation, for which the best results are obtained applying methodologies of co-engineering and co-design.

18.1.4 Trend of Evolution for the Production Means

We have seen that the requests of machine tool users are focused on the final results of the job, but we must consider that other constraints still exist. In many cases they are not explicit but concur in the final choice of the production means.

First of all we have to consider the general performance of the machine; that is still of great importance and absolutely cannot be neglected; the risk is not being able to justify our product in comparison with competitors. Besides we should not forget that in many cases the new organizing solutions are possible only if we utilize the newest technical solutions for the plant components; we can mention, for example, transfer lines for the automotive sector: multi-spindle stations can be substituted with milling modules only after the advent of high-speed machining.

To complete the final choice we have also to consider other criteria, driven by legal issues or ethical criteria. These are for example the safety requirements or the ecological compatibilities, that very seldom enter into the requirements clearly expressed by the customers, but that are important assumptions for the final choice.

18.1.4.1
Lean and Agile

Lean and agile are two very fashionable adjectives to define the organization models for the new productive structures, and by analogy to define the machines that must operate inside these environments. Until not very long ago the concept of a "lean" factory seemed partially in contradiction with the concept of "agile," promoted by a further methodological approach. Now the two models are much more clear, and today the dominant idea is that "lean" is the fundamental requirement, a basic characteristic to evolve towards agility. Agility cannot be an objective to pursue by itself, otherwise as a result we can obtain a production structure that is not prepared to follow the evolution of products and technologies, a structure that will not be able to be updated, but be subject to a rapid obsolescence.

The situation is very different instead if we consider "leanness" as a requirement to define an agile machine; eventually our machine will have a higher initial cost, but will be provided with the dose of flexibility and reconfigurability necessary to follow the future products' evolution.

18.1.4.2
Flexibility and Reconfigurability

Solutions based on hard automation, the so-called automated factories, are practically dead. These solutions in practical cases demonstrated not to be able to support the product evolution and to be technically updatable only at very high costs of reprogramming and reconfiguration. It was instead demonstrated that the physical presence of an operator could be an important element of flexibility for the productive systems without reducing productivity, if the human element would not introduce causes of variance for the machining time.

A second theme substantiated by a great number of research activities all around the word, is related to modularity, or better functional modularity, on which will be based the new generation of production machines and systems. According to this approach, systems are based on elements, the functional modules, that can perform in an independent way the operations for which they have been designed, and can be easily plugged-in and connected to the production system.

Following this approach the design of machines is rapidly changing: working units with high productivity levels but rigid, are definitively losing in comparison to more flexible units, even if the latter are a little less productive. We have important examples in the automotive sector, where very fast single-spindle working units often are substituted with multi-spindle units. The latter have a lower production capability in respect to old solutions, but they can be easily reprogrammed to follow the requirements of the production.

18.1.4.3
Modular Machines and Plants

From the point of view of machine tool manufacturers, a modular concept of machines and plants seems to be the only possible solution to satisfy those customers that require customized solutions and high levels of performance, but are not willing to accept penalties in terms of reliability and costs. The modular approach allows machine tool manufacturers to specialize inside their own competence field, ideally becoming "best in class" for this matter, and participating with other colleagues in the construction of the final system.

Obviously such a kind of modular approach has very strong implications on the machine conception and on the automation models that it can be applied to. Inside this vision, standardization of mechanical, electric, and electronic interfaces will be a most important argument. Another challenge that manufacturers of electronic apparatuses will have to face in the nearby future will be related to a larger distribution of control devices along the production lines and consequently to the need of simple connections of all the elements that compose the system.

18.1.4.4
Reliability

The term *reliability* is commonly used, even if the most proper term should be availability, a parameter that involves also data related to maintainability, i.e., the possibility to individuate damage and repair in a very fast way. Maintainability is an index that defines for how much time a machine is in an operable condition.

Increasing availability is an objective that should involve the entire company structure of the manufacturer, from the design department to the on-field service department.

We have to outline that in terms of availability, performances of a production means depend greatly on the initial phase of design. Henry Ford already understood this concept while defining the famous 'model T' car: more parts are in there, so more parts will crash. Also in this field, 'lean' concepts are giving a strong impetus to simplification, to a reduction of redundancy, and to a less protective dimensioning.

A second fundamental aspect is the reliability of components. On almost every machine built in Europe, at least 30% of the total value is composed of outsourced components. Approaches to the design based on methodologies, such as Fault Tree Analysis, that allow a prediction of the final level of reliability, are not yet widely used, but they should be useful to make a step forward, similar to what electronic components did: numerical controls in the past used to be a major source of problems for industrial applications; today they are one of the most reliable parts of a machine tool.

To increase the efficiency of service, electronics can be of great help: today it is possible to connect via modem the Computer Numerical Control (CNC) of a machine installed anywhere in the world, with the location of the manufacturer, and this possibility can be used for tele-diagnosis or teleservice. In many cases it is possible to find the cause of malfunctioning, in some cases it is even possible to fix it (mainly for software errors). At a minimum, it is possible to have a clearer idea on the situation and to organize a more effective service intervention; in the easiest cases it is possible to give instructions to the customer for a direct repair.

Using the possibility of connection offered by the Internet, it is also possible to share with the customer updated information or to allow service people to access technical databases to get all the information needed for adjusting the machine.

A main obstacle to a wider diffusion of these systems is that commercial CNC are not studied for this application, and they have a reduced possibility to exchange data with Programmable Logic Controller (PLC). Besides, on-board diagnostic systems are not ready to fully accept these functions, but it is easy to foresee that in the near future teleservice will become a very widespread activity.

18.1.4.5
Productivity

Leanness is also a way to go faster. It is estimated that with traditional design methods, on every machine 50% more power is installed than is strictly necessary, and this is partially related to the over-dimensioning of structures. It is clear that a 'lean' approach is not compatible with 'fat' machines, and this is more evident for high-speed machines; it is not possible to reach top performance without a reduction of inertia, i.e., the weight of moving elements.

To move very fast axes, linear motors are being considered and appreciated today, mainly for lighter applications. Linear motors are solutions intrinsically lean; they allow a drastic reduction of all the mechanical elements that compose the kinematic chain, and the number of their applications is rapidly increasing.

Any kind of actuator we decide to use, the objective is to increase the dynamic performances of the machines, or, to express this concept more properly, to reduce machining times. High velocity can be a little confusing definition: in the common sense there is no distinction between high cutting speed (or spindle rotation), high cutting feed (that involves the possibility to control the trajectories in the space), and rapid feed (that primarily relates to the acceleration).

In a machine of the new generation we should tend to machine 'quickly,' instead of at 'high speed.' And quickness can be reached with different criteria according to the application to which we refer: speed for working aluminum, feed for finishing a mold, or acceleration for a machining center with small travel lengths. The concept can also be referred to different machines that today are not usually considered for this kind of application: why not also consider, for example, presses and their number of strokes per minute?

18.1.5 Two Case Histories to Think About

How and how much investing on leanness, where and how to introduce flexibility are decisions that must be taken, looking in a different way for each single firm, considering its own production and strategies. Two cases of very different firms can help us in making some considerations.

18.1.5.1 Case 1

A very large company, manufacturing automotive components, decided to start a new production facility in a foreign country. Initially they started the production of 'Type A' parts, but already were foreseeing to substitute them with 'Type B' products in the next three years. Type B would have better performances, but the production method would be quite different.

The company manages, and actually owns, all the technologies for their own products, and it is prepared to make autonomous market projections on which investments are based.

Figure 18.2 shows the production program and indicates the startup cycle of 'Type A,' initially in an experimental way, to reach the budgeted volume with definitive production lines.

While this is running, it will introduce 'Type B,' whose commercialization is foreseen after three years, but for which it is mandatory to use as much as possible of the existing production plant, partially reconditioned. Production of Type B should evolve in the future, according to the estimation of market response, to very high production levels.

It is clear that this manufacturer will need two different plants, but they should be easily reconfigured to follow the different requests of production. The test line should be conditioned to accept both versions, which means it will be very flexible and reprogrammable. The series-production line should be modular and reconfigurable. The level of flexibility, meaning the capacity to adapt in a fast way to the new production, will mostly depend on the proper evaluations of market trends (Is it true that model A will be out of market? How long will be the transition period between A and B?), and on internal evaluations of the minimum batches judged economically acceptable.

18.1.5.2
Case 2

The second case we should like to mention is relative to a manufacturer of large machine tools, i.e., machines that are subject to continuous changing and customization: where great flexibility is of utmost importance.

For this application, the prosecution of a lean manufacturing structure goes through the research of greater efficiency. The need of flexible production machines in the workshop is a must, no other solution is possible. So, the intervention focused on two different points:

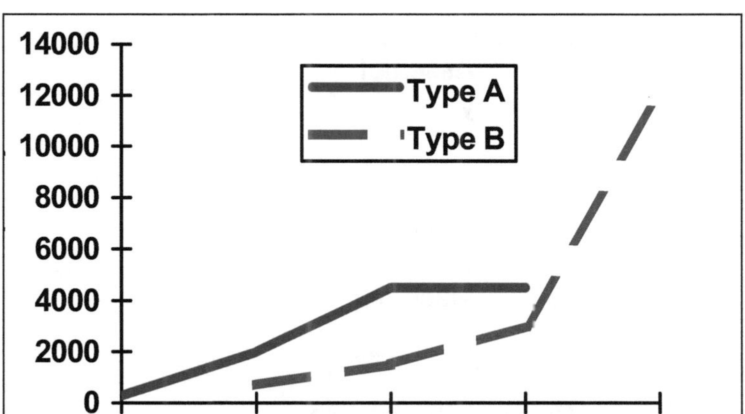

Fig. 18.2 Two different products must be produced in two different periods, optimizing the investment.

– The use of the critical production means must be optimized;
– Assembly times must be reduced.

To fulfill the first request, the company posed the objective of reaching 700 hours/month of machining time with just two manned shifts. And this led to big investments in automation of the machines and on the level of reliability.

To fulfill the second request the company posed a new objective: during the assembly phase workers cannot adjust mechanical parts, they can only assemble. All machining operations must be performed in the workshop, even the smallest ones like bevelling or making small tapered holes, operations that traditionally are performed while mounting the machine. To reach this objective, the better utilization of production means reduced the cost per hour, allowing workers to perform also those operations that initially would have appeared less economically convenient.

The main advantage of this policy has been the dramatic reduction of assembly time (see Fig. 18.3): in 1960, it represented 80% of the total production time for a machine, in 1995 it was reduced to 30%. This approach led also to the reduction of the variability of assembly times, that everywhere else is the most important cause of delay in the delivery of the final product.

In this case the company followed objectives of efficiency and agility, improving a structure that had to remain flexible, investing in automation for the mechanical production, and implementing more rational and efficient procedures for the assembly department.

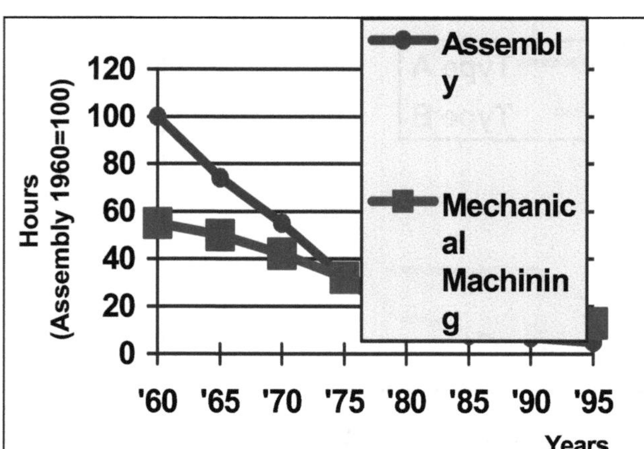

Fig. 18.3 Assembly time has been reduced in absolute value in contrast to mechanical machining time.

18.1.6
Trendy Technical Topics

18.1.6.1
Dry and Semi-dry Machining

Dry and semi-dry machining are deeply studied inside research institutes and began to find the first industrial applications mainly in Europe and Japan. Adoption of these techniques is today very limited: for an effective use of dry or semi-dry machining, there are needed machines specifically studied for a fast evacuation of the hot chips that the machining produces, without the help of a coolant; in many cases it seems realistic that a possibility exists to join a better social compatibility of the process (less pollution, safer environment) to a reduction of machining costs. (The cost of coolant and of all the connected devices represents roughly 15% of the total machining cost.)

In these days we observe that operations that require pure cutting oil, such us gear cutting, are progressively being substituted by dry cutting.

For more traditional operations like drilling, milling, and turning, new tools are being developed, that allow the abolition or at least a dramatic reduction of the quantity of coolant used for their operation. With MQL – Minimum Quantity Lubrication – techniques it is possible to reduce the amount of coolant by many orders of magnitude: it moves from 10÷20 liters/min to 20÷50 milliliters/hour. Using this technology, coolant evaporates in contact with the part and the chips. It is necessary to use environmental friendly oils, typically vegetable oils, and this is a further improvement towards ecological machining.

18.1.6.2
High Velocity

The increase of dynamic parameters of machine tools (speed, feed, rapid) today is one of the most important arguments connected with the evolution of machine tools, also for commercial or promotional purposes.

The dynamic performance, that presents the major influence in the definition of the costs of production, changes accordingly to the application considered, and consequently the solution proposed must change.

If we analyze three different applications, typical for three different sectors, we can better understand how high velocity can be utilized in different ways. As examples, we can consider the machining of aeronautical body components, the finishing of a mold, and the machining of automotive components such as gearboxes.

In case of aeronautic components, where it is rather common to cut away 90% of the initial material, the critical parameter is the volume of chips produced in a unit of time. Materials to be machined are typically aluminium alloys, and the customer requirements are for high cutting speed to obtain high cutting feeds. Today tools exist that are very resistant to high cutting speeds, and in many cases the limiting factor must be searched inside the machine tools that do not have enough power and dynamic characteristics to use the tools to their fullest potentiality.

Remaining in the field of aeronautic applications, from the point of view of materials representing the new frontier of mechanical machining, there exist other applications for which the possibility to have more performing tools should be much appreciated (see Fig. 18.4): this is the case of the manufacturing of titanium alloys, which are very tough to machine, and composite materials, for which the reinforcing fibers can be very abrasive. Cutting tools for these applications can be improved, and this is one of the developing areas for the future.

Materials for mold and dies are typically cast iron or carbon steel. Most innovations today regard finishing operations: for these applications, the limiting factor was the possibility of controlling tool trajectories in the space with high feed values. With last-generation CNC, this aspect is no longer critical, and now the limits have moved to the machines and their dynamic characteristics, and to the tools and their resistance to wear, necessary to improve the economy and the precision of operations.

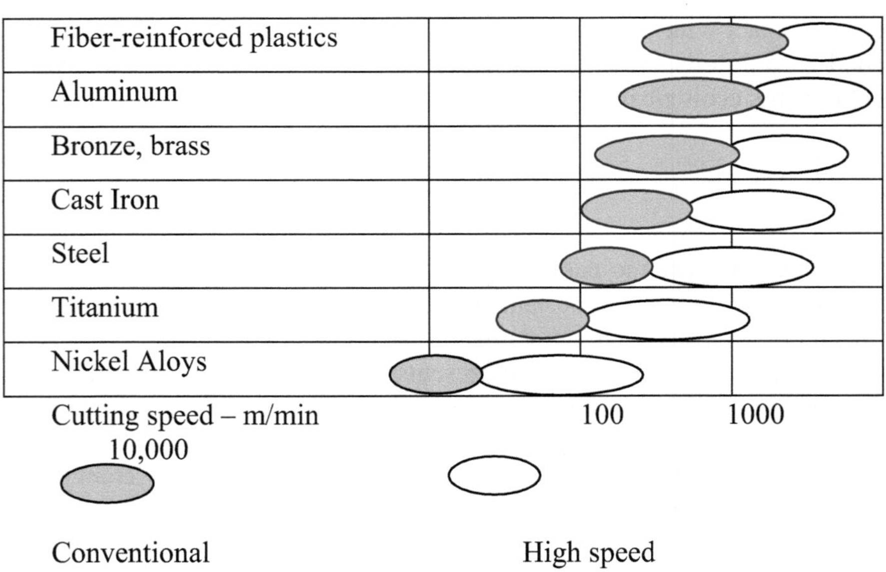

Fig. 18.4 Definition of high cutting speed depends on the application.

The case of automotive operations is even more different. The adoption of production lines with the innovative architecture mentioned above, becomes possible only with the realization of milling modules with innovative characteristics, mainly in terms of velocity and acceleration, necessary to reduce idle time and particularly reduce the times spent at rapid speed to move the tool from one working area to another.

For all these applications, it was necessary to re-define the machines from a dynamic point of view. When acceleration increases, moving masses become a critical element: in many applications, forces of inertia are higher than the cutting forces.

This process of re-designing machines brought astonishing results. Using classical architectures it was possible to increase performance from 2 to 5 times, while to change the order of magnitude of kinematic parameters (to increase them 10 times), it was necessary to completely change the concepts used in the past, and parallel kinematic structures, the so called hexapods, were introduced.

For machine tools, this is a real change in terms of reference that opened a scenario of completely new solutions. Today the first real parallel kinematic machines are beginning to enter the market, solutions that focus on specific applications for which parallel kinematics can introduce the best advantages. Rigidity is one of the intrinsic limitations of this kind of structure; even if the calculated rigidity is high, in the real case it is not easy to match the theoretical results, on the other side, the best advantage is the reduction of masses that allows fantastic dynamic performances. These aspects suggest a good field of application where operations need high speed but forces are not too high, such as finishing operations or handling.

High-speed machines allowed new configurations for the production lines, for which concepts of leanness and agility are a must. The new approach foresees mainly the utilization of single-spindle units, instead of the old multi-spindle units that are more rigid and less re-configurable in following changes in the production mix. It is clear that new organization models can be introduced only if there is no loss in production; they can be realized only because new technologies are on the market: new machines must be fast and quick to reduce idle times, tools must be resistant to wear to last longer at higher cutting speeds, and all the components of the machine must guarantee performance adequate to the new expectations.

18.1.6.3
Hard Machining

In the development of the overall economy of machining operations, high velocity is not the only way that can be pursued. One very encouraging field of evolution (see Fig. 18.5) is the reduction of the number of phases and number of machines necessary to obtain the desired part. Hard milling and hard turning are following this philosophy, they are cutting operations on hard materials, typically tempered, that today are finding the first industrial applications. These new technologies are extremely promising in reducing the cost of machining, but they have a strong impact on the machines. These must be rigid as cutting forces are high and precise, since they must perform finishing operations substituting for grinders. Hard machining also has a strong impact on tools that actually are the key point for a real application: they are an enabling technology; this kind of operation becomes possible only if the right tools are on the market. Today this is still a critical aspect, and profound knowledge of tools is an argument of competition for those who propose these applications.

18.1.7
Conclusions

The buying process of production means is changing; customers are ever more interested in the final results of the operations and not in which way they are obtained. The role of suppliers is changing; they must be more like technical partners than sales people.

Machine tools are rapidly evolving in terms of performance and of organizing solutions for the workshop. New production lines or organizing strategies are possible only in front of new machines.

High velocity, dry machining, hard machining, and modularity today are the key arguments for the development of machine tools.

18.2
Intelligent Machine Tools

18.2.1
Necessity of Global Intelligent Manufacturing in Modern Societies

T. Nagao, Y. Hatamura, M. Mitsuishi, M. Nakao

Strong demands are in place on modern manufacturing systems, especially on those machining systems for cutting and grinding. These demands in clude high efficiency, high accuracy, conformance to information systems, and low impact on the environment. The authors have introduced intelligence into manufacturing systems to satisfy two of these needs, namely higher accuracy and support of information systems.

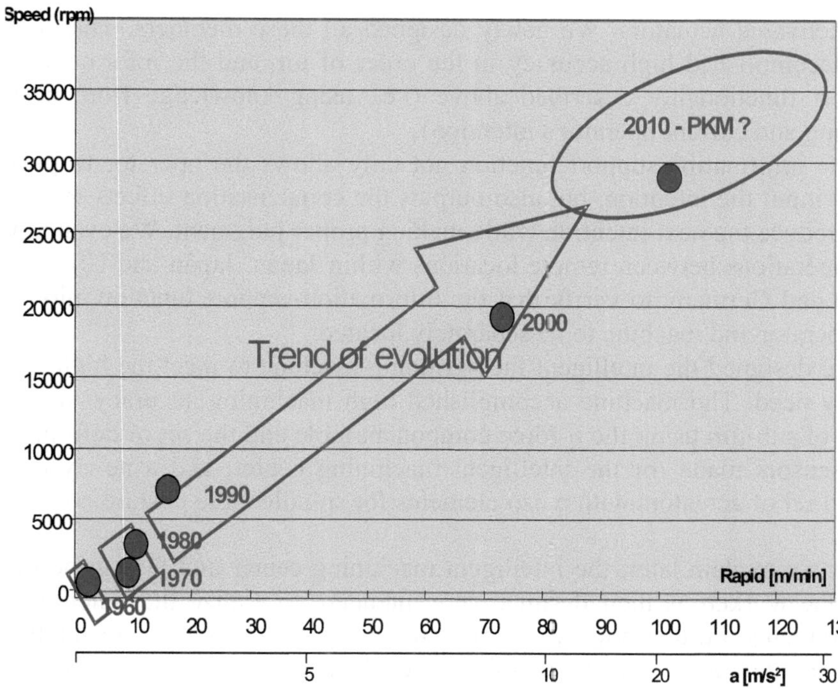

Fig. 18.5 Evolution of dynamic performance of machine tools.

This intelligent manufacturing system, as Fig. 18.6 shows, has a base structure of new sensors, controllers, and actuators mounted on a conventional machining tool to realize the following operations: Sensors detect physical phenomena during actual machining in real time, then controllers compare and analyze the prediction of next physical phenomena based on the sensed values with past physical studies, knowledge, and current operator's intention, and lastly actuators produce desired functions compensating for the control parameters. The system again detects the physical phenomena after correction to evaluate the effects of its compensation, and this way, the controller acquires new knowledge to gradually increase its intelligence level.

The authors, for about 20 years, have developed these sensors, controllers, and actuators to give intelligence to conventional machines. The next subsections 18.2.2 and 18.2.3 describe an intelligent machining center and an intelligent face-grinding machine which are examples of such intelligent manufacturing systems.

We designed the intelligent machining center to meet the need for higher accuracy and information-system support. The system structure includes a 6-force component table, a set of deformation sensors, and a set of heat-activated actuators. We newly designed all these members. The system accomplished high accuracy in the order of μm and the information-support functionality described above (i.e., using knowledge from past learning and current operator's intention).

The information-support function not only allows the operator to precisely input the intention, but also outputs the compensation effects so she can produce the next intention while making proper judgment. We even ran tele-operations between remote locations within Japan, Japan and US, and Japan and Germany to verify that the information-support function allows the operator and machine to be separately located.

We designed the intelligent face-grinding machine to meet the high accuracy need. The machine accomplished high machining accuracy in the order of sub-μm using the 6-force component table and the set of deformation sensors made for the intelligent machining center, and a newly designed set of actuators with piezo elements for spindle/table posture correction.

As we explain later, the intelligent machining center and face-grinding machine worked as their designs were intended, to realize the functional requirements. Furthermore, we applied the design methodology for intelligent manufacturing systems to solve other difficult processes like drilling holes on printed circuit boards, cutting photosensitive drums for copiers, and grinding complex mold surfaces. We are currently extending our development to master-slave systems for remote surgery and a system for cutting and assembling micro-machines using electron microscopes. Through our development, we found what follows:

a) Satisfying the requirements for high accuracy and information support requires "non-control components" separate from the above-described control system. An example of such components is fail-safe mechanism hardware for the intelligent machining center. Relying on control alone can damage valuable systems (especially the sensor) upon an abnormal operation, e.g., a system that depends on information processing of an excessive load signal for avoiding accidents. Also, the intelligent face-grinding machine uses a non-deforming plaster chuck. Fixing a thin plate without this chuck changes the plate's flatness, once released from the chuck after ground perfectly flat, due to elastic deformation from the strong chucking force springing back. Such elaborate schemes are possible like measuring the plate deformation before grinding and controlling the grinding depth according to the elastic-deformation estimate, however, the grinding takes too long, clogging the grinding wheel and eventually hindering proper control from changes in the grinding efficiency. A good approach is not to incorporate all components into the control system, but to keep some purely mechanical functional hardware that are not in the control system.

b) To satisfy the requirement for high accuracy machining, physical phenomena should have "repeatability" during past learning and current control periods. If not repeatable, what acquired during the learning period is useless for control. Examples of such phenomena include the following: The intelligent face-grinding machine had shims between faces of mechanical contact to meet the assembly tolerance. They caused hysteresis in the actuator compensation. The machine used pre-loaded linear ball-bearing type infinite guide tracks which occasionally generated a max. 0.3μm vibration when the balls are pressed in the direction to increase the pre-load. Compensating for such vibration is only possible if it repeatedly makes the same amplitude. Hardware that guarantee repeatability are minimum requirements for intelligent manufacturing systems.

c) Figure 18.6 shows the feedback control-system construction with knowledge. As the "time delay" increases, the control does not converge. Examples of coping with time delay include the following: For the intelligent machining center, when the tool chips or causes vibration, notifying the operator about the failure to receive instructions delays the response, and so we set the logic for the machine to immediately make a judgment and set the tool feed smaller. In case of the intelligent face-grinding machine, when it vibrates, as mentioned before, detecting the vibration from the ground surface waves is too late for any

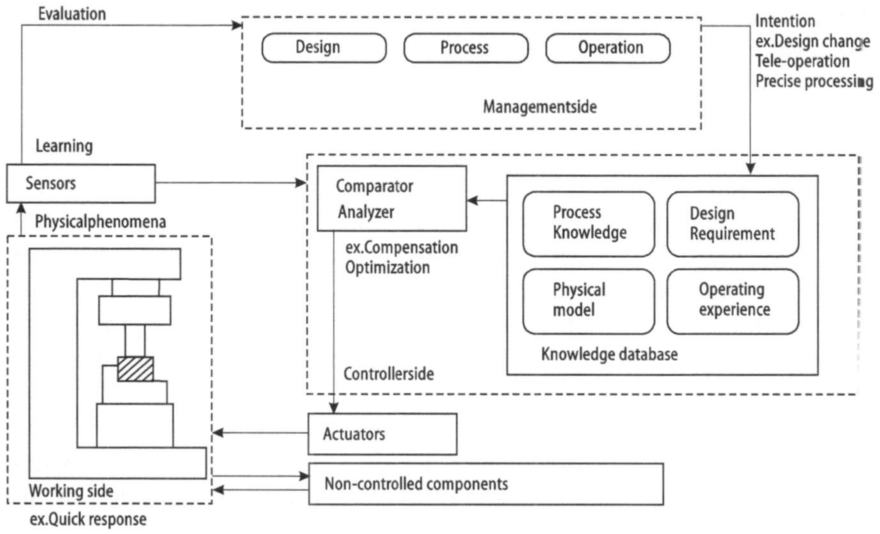

Fig. 18.6 Structure of intelligent manufacturing system.

countermeasure, so we placed a reference surface next to the work and detected the vibration component in real time to control the table posture.

The authors faced these design obstacles, and through some trial and error, established design methodology for intelligent manufacturing systems as we describe in subsection 18.2.2.

18.2.2 Development of an Intelligent Machining Center

M. Mitsuishi, Y. Hatamura, T. Nagao

18.2.2.1 Introduction

The authors have developed a high-efficiency machining system, which maintains a stable cutting state using multi-axis force information from sensors installed on the table of a machining center. The implemented system has the capability for cutting state monitoring, adaptive control by modifying the spindle speed and the feed speed in real-time, and learning through the storage of an *experimentally determined stability-lobe diagram* which incorporates knowledge representing the relation between machining conditions and the machining state. This article discusses the following items: (a) machine tools with open-architecture CNCs, (b) real-time machining-state detection, (c) an efficient stability-lobe acquisition method as a practical database, and (d) a database-sharing method among different machine tools.

Concerning the thermal deformation of the machining center, the thermal distortion was actively compensated by thermal actuators, consisting of heaters and coolers which are located in the lower part of the column. The translational error was compensated by shifting the origin of the coordinate system of the numerical controller, and the rotational error was compensated by controlling the thermal actuators. The amount of shifting and controlling was determined by a neural network system integrated with a genetic algorithm and a neural network system with the feature of 3-state sigmoid function. Using this method, although the room temperature was varied during the experiment over a range of 14°C, the position and the attitude of the spindle were maintained within 12μm and 21.4μrad, respectively. The achieved result was approximately 1/3 of the variation of the uncontrolled machine.

Global manufacturing is becoming increasingly feasible as information networks such as computer networks become widespread. This fact should drive basic changes in methods for technology transfer and education, as well as providing a means for the remote operation of machine tools. The manufacturing system developed in the article incorporates a reality-transmission capability, which provides an environment for the simultaneous transfer of know-how and experience without the restrictions imposed by place and time. The technologies discussed in the article can be used to establish a remote operating and monitoring system with an advanced user interface that can be used for routine production systems in the future.

18.2.2.2
System Construction

Figure 18.7 shows an overview of the hardware system described in this article. A machining center, which is controllable from an external, real-time computer developed by the authors' group, was used. The machining center was equipped with a 6-axis force sensor to monitor the cutting state and a fail-safe table to prevent damage to the system in case, for example, of an over-force condition. The rigidity, the natural frequency, and the rated values of the 6-axis force sensing table developed by the authors group are approximately 500N/μm, 1kHz, and 2,000N, for the spindle rotation direction and 1,000N for the orthogonal directions, respectively. The output signal from the force-sensing table is amplified by a strain amplifier

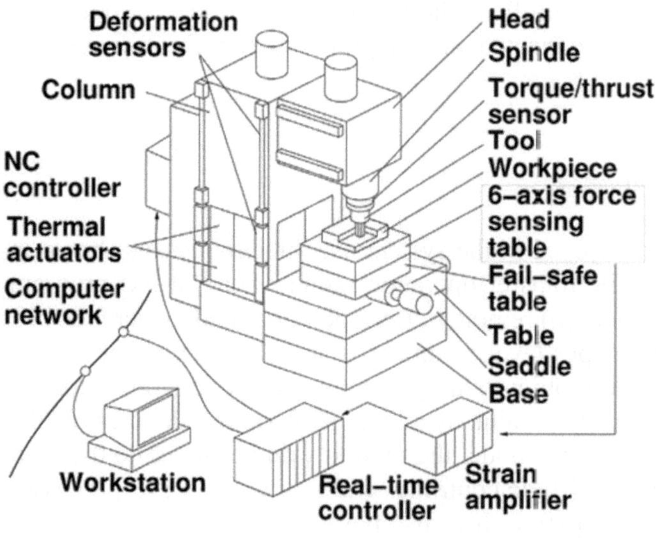

Fig. 18.7 Hardware system Overview.

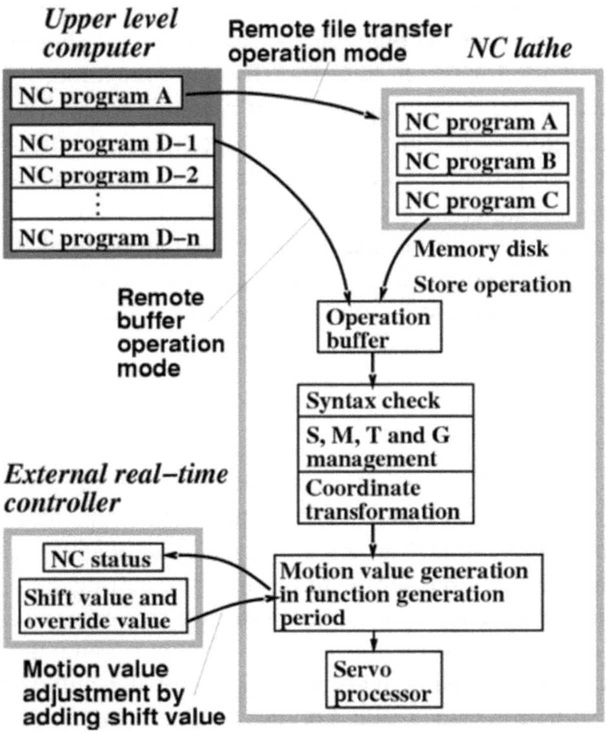

Fig. 18.8 CNC controller.

and sent to the real-time controller. The real-time controller, the database, and the CAD system are connected and integrated with each other through a computer network.

18.2.2.3 *Open-Architecture CNC*

The CNC controller developed in the system has three modes, as shown in Fig. 18.8: (a) Remote file transfer operation mode where an NC program can be transmitted at once, (b) remote buffer operation mode where the NC program sent from the upper-level computer is stored in the buffer and simultaneously sent to the motion-value generator, and (c) adaptive control mode. Modification of motion value by adding shift value characterizes the adaptive control mode. In this mode, the motion value is added to the shift value from the external real-time controller and sent to the servo-processor every function-generation period. It is also possible to change the override value of the CNC controller by sending an override value from the external controller. Furthermore, the open-architecture CNC controller has the function of transmitting the NC status upon the request of the external real-time controller.

The external real-time controller and the CNC controller were connected through Ethernet. Furthermore, UDP/IP was adopted for a lower level communication protocol, because of its high communication speed. Reliable communication was secured by preparing a simple handshaking protocol. Following the command from the external real-time controller, the CNC controller replies wi4h the status of the machine tool. The frame check block and packet sequence number are added, and the receiver checks for disorder. Using these techniques, the round-trip communication time required was approximately 5 ms.

A controller, which enables remote manufacturing by communicating interactively with another computer during machining, needs to be equipped with a function to accept information for the servo-controller, for example, at the function-generation level. Furthermore, the controller should be connected to an information network, such as the Internet, to send the sensor information and the controller's state to the remote operator when it is requested. Conventionally, the goal of an open NC controller is to provide an interface that is defined clearly at several levels in the NC controller. Following this tendency, transparency through the network is also desirable and is becoming available. However, the systems developed by other researchers only have the capability to display a control panel on a WWW browser connected to the network, which allows the user to remotely operate the system. Compared with these systems, the system developed by the authors has a capability to operate the machine tool as if he/she were holding a cutting tool by themselves. An external real-time controller as prepared to send the information from a remote operator through the network to the NC controller with an open interface [1].

18.2.2.4
Fail-safe Mechanism

In the proposed machining system, a human operator controls the system. It is impossible to eliminate human error completely. Furthermore, it is difficult to predict abnormal cutting conditions when the geometrical shape of a workpiece is complicated. Therefore, the system must be designed such that the system is not damaged even if excessive force is applied to the system. "Fail-safe capability" is crucial to realize such a system. The necessary functions of the fail-safe mechanism is as follows: (1) High rigidity should be maintained in the normal state. (2) Large displacement occurs when an excessive force is applied to the system and the fail-safe system absorbs this force. (3) An emergency signal is generated when excessive force occurs and the system is stopped. (4) When the excessive force is removed, the system returns to its normal state. To implement the above-mentioned functions, tensile and compression springs are combined. In our system, the fail-safe mechanism was realized using leaf-springs.

18.2.2.5
Software System

Overview

Figure 18.9 shows an overview of the software system. The system consists of a CAD system, a database, and a real-time controller. One of the features of the system is that these processes are not static, but dynamic, communicating with each other during the cutting sequence. More concretely, the contents of the database are dynamically revised by being sent the actual cutting conditions and cutting state from the real-time controller. Furthermore, the three-dimensional modeler in the CAD system can be accessed dynamically during cutting. Tool path and cutting conditions are regenerated depending on the revision of the database and are sent to the real-time controller.

Database

The database consists of an "experimentally obtained stability-lobe diagram," a "machining-efficiency-evaluation task," and a "machining-condition-determination task." The stability-lobe diagram introduced by Mer-ritt [2] and Tlusty is experimentally obtained in the system described in this article. Therefore, it is called an "experimentally obtained stability-lobe diagram."

The machining-efficiency-evaluation task and the machining-condition-determination task use the 3-dimensional modeler in the CAD system to determine efficient machining conditions and generate an appropriate tool path. Furthermore, the machining-efficiency-evaluation task modifies the stability-lobe diagram as it varies during machining.

Real-time control

The "NC-data-interpreter task" acts as an interface between the CAD system and the real-time controller. The NC-data-interpreter task accepts any arbitrary NC data set which is described in the APT format. The "real-time machining-condition controller" and the "real-time machining-state-judgment task"[3] perform adaptive control by adjusting the override values for spindle speed and feed speed based on the previously determined machin-

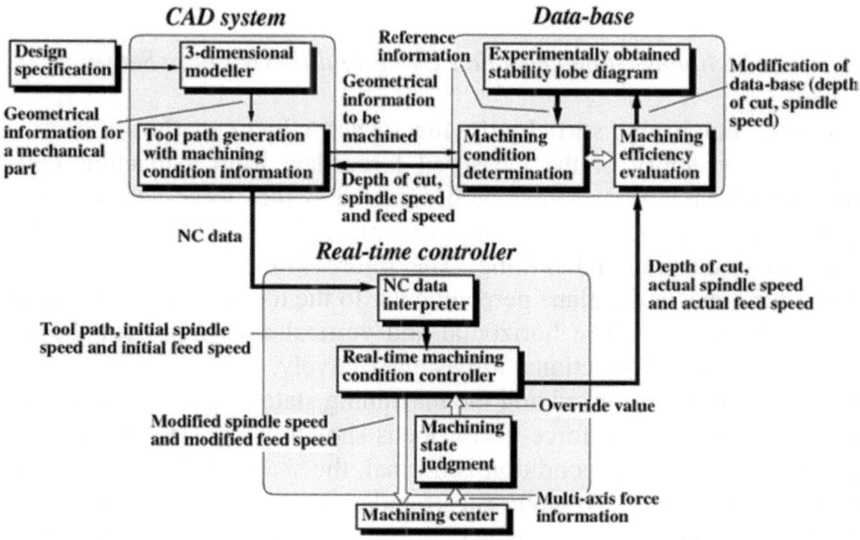

Fig. 18.9 An overview of the software system.

ing conditions. Furthermore, the relationship between the machining conditions and the machining state is fed back to the database.

CAD system

In the CAD system, geometrical information for the component mechanical parts of a product is input into the 3-D modeler and the abstracted machining-feature information is sent to the machining-condition-determination task in the database. The stability of the end-milling system depends on the spindle speed and the depth of cut, according to the stability-lobe-diagram theory [2]. Therefore, the machining-conditions-determination task returns the most efficient set of spindle speed and depth of cut for the given machining geometrical features. The tool-path-generation task generates a 3-dimensional tool path appropriate to the received depth of cut and geometrical machining-feature information.

The most efficient set of spindle speed and depth of cut obtained from the machining-condition-determination task changes when the stability-lobe diagram in the database is modified. Therefore, in the system described in this article, the tool-path-generation task is invoked periodically during machining, and the tool path and cutting conditions are quasi-dynamically modified to maintain maximum cutting efficiency. Typically, the tool path is regenerated after each cutting pass.

18.2.2.6
Real-Time Machining-State Detection Using Multi-Axis Force Sensing

Judging the machining state by the force-vector-locus method
The plotted trajectory of the measured force data from a machine-tool-mounted, multi-axis force sensor, is designated as the "force-vector locus" by the authors [3]. Figure 18.10 shows an example of the force-vector locus in full immersion-end milling. The figure shows the 2-dimensional force-vector locus in the plane perpendicular to the tool-rotation axis, using time as a parameter. The horizontal and vertical axes represent the X-directional and the Y-directional forces, respectively.
The proposed method for judging the machining state may be explained by taking the 2-dimensional force-vector locus shown in Fig. 18.10 as an example. When the cutting condition is normal, the shape of the force-vector locus resembles a circle. In contrast, irregular variations are superimposed on the normal force-vector locus under conditions of high-frequency chatter. Chatter results from the resonant vibration of the cutting tool, which is excited by the cutting process. Furthermore, the shape of the force-vector locus becomes flattened, like an ellipse, when beating occurs. To evaluate the shape of the force-vector locus, the following calculation is executed: the center of gravity of the area bounded by the closed locus is calculated for each tool rotation and the distance from the center of gravity to each subsequent sample point is calculated. The variance of these distances is calculated subsequently. The machining state is judged to be normal if the variance is small, and chatter is deemed to have occurred if the variance is large.

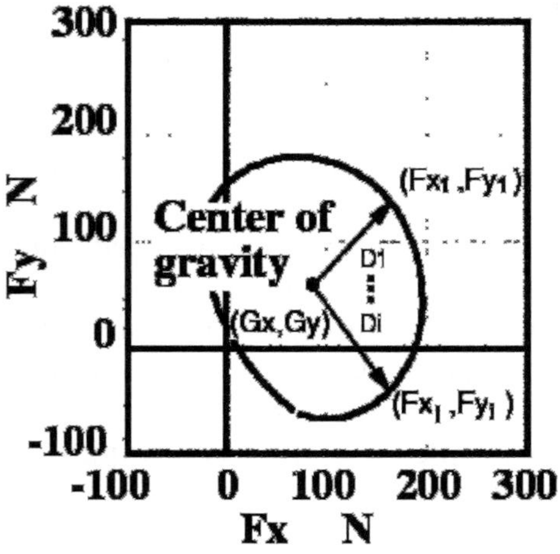

Fig. 18.10 An example of the force vector locus in full immersion end milling.

Theoretical analysis of cutting force

As has already been mentioned in the previous section, the variance of distance varies with the machining state. However, the variance also varies with the cutting conditions, as well as the machining state. That is, the variance of the distance is larger under conditions which generate high cutting forces, as compared to those which produce small cutting forces. Therefore a threshold variance must be determined to accurately determine the actual machining state. Since the experimental determination of a threshold in each case would eliminate any benefit of the system, the threshold variance is calculated through the use of a physical model of the cutting process. In this article, a theoretical analysis of square-end mill cutting is described.

It is assumed that the cutting force may be obtained by multiplying the normal cross-sectional area of cut by the cutting force per unit area of the cut, *kc*, according to the Kronenberg equation. *kc* is represented by using the undeformed chip thickness, *h*, considering the size effect.

$$k_c = Ch^{-\varepsilon} \tag{18.1}$$

where C and ε are constants which depend on the workpiece material and, to a lesser extent, on the shape of the cutter. The total cutting force and the cutting force components in each cutting direction can be calculated, as follows, by integrating the infinitesimal cutting force from the rotational angle at the start of cutting, $\phi\ \phi a$, to the rotational angle at the end of cutting, ϕ ϕb [3].

$$F = \int_{\phi_a}^{\phi_b} \mathrm{d}F = \frac{rC}{\sin\gamma} \int_{\phi_a}^{\phi_b} h^{1-\varepsilon} \mathrm{d}\phi \tag{18.2}$$

$$F_x = \int_{\phi_a}^{\phi_b} \mathrm{d}F_x = \frac{rC}{\sin\gamma} \int_{\phi_a}^{\phi_b} h^{1-\varepsilon} \cos\Psi \sin(\phi-\theta)\mathrm{d}\phi \tag{18.3}$$

$$F_y = \int_{\phi_a}^{\phi_b} \mathrm{d}F_y = \frac{rC}{\sin\gamma} \int_{\phi_a}^{\phi_b} h^{1-\varepsilon} \cos\Psi \cos(\phi-\theta)\mathrm{d}\phi \tag{18.4}$$

$$F_z = \int_{\phi_a}^{\phi_b} \mathrm{d}F_z = \frac{rC}{\sin\gamma} \int_{\phi_a}^{\phi_b} h^{1-\varepsilon} \sin\Psi \mathrm{d}\phi \tag{18.5}$$

where γ is the cutter helix angle. In these cutting force calculations, the following parameters are unknown and defined, as follows:

C: proportional constant of the cutting force,
ε: exponential coefficient of the cutting force,
ψ: angle between the infinitesimal cutting force dF and the plane which is defined by the infinitesimal cutting forces dFx and dFy, and
θ: angle between the infinitesimal radial component of the cutting force dFt and the resultant force of dFx and dFy.

Since the integration mentioned above cannot be performed analytically to calculate the unknown parameters, the Gauss-Legendre integral formula was utilized for evaluation of the integral. C and ε were determined using the method of extremum, using a penalty function equal to the square of the difference between the experimental value and the calculated one. ψ and θ are determined by experiment, where θ is calculated as the linear approximation of ϕ. It should be noted that the parameters must be determined from a sufficient number of experiments under different machining conditions.

Experiments

Normal cutting-state data and chatter cutting-state data are compared in full immersion square-end milling. The machining conditions were 2,200 min^{-1} spindle rotational speed, 1.0 mm/sec feed rate and 2.5 mm depth of cut with a 5 mm diameter, square-end mill with axial angle $\pi/6$ rad. The machining was performed in the Y direction, on 0.45% carbon, plain carbon steel. A high-frequency vibration is superimposed on the cutting force signal under conditions of chatter.

In the Fourier-transformed data, a peak can be seen at about 800 Hz which is caused by the chatter. However, the comparison of both the crude force data and the Fourier-transformed data are not easily automated, and a visual search of the data is time-consuming for real-time supervision. Furthermore, several tool rotations are required to generate the Fourier spectrum.

In the following discussions, the proposed machining-state judging method is applied. Force data during one tool rotation were plotted as a force-vector locus. When the machining state is normal, the shape of force-vector locus is similar to that calculated using the physical model. In the chatter state, the shape of force-vector locus is flat, and irregular, high-frequency variations are superimposed. From the data which present the time change of the variance of distance in force-vector locus when the machining states are normal and under conditions of chatter, it can be seen that the variance of distance in the force-vector locus under normal machining is close to that calculated using the physical model. On the contrary, in the chatter state, the variance of distance is large compared to that calculated using the physical model. Furthermore, the variance itself varies significantly as a function of time in case of chatter.

From the experimental results, the machining state can be judged efficiently by comparing the measured variance of distance in the force-vector locus and that calculated using the physical model.

Summary

From the method described in this article and experimentation, the following conclusions can be made:

1. When the machining state is judged using the Fourier-transformation method, thousands of data samples from a multi-axis force sensor are necessary. In case of digital data processing of 12-bit data, 2^{12} of data is necessary for sufficient accuracy. For example, when data is sampled at a 2-kHz rate, data collection over a period of at least 2 s is required. Therefore, Fourier-transformation methods are too slow for real-time supervision of machining. Furthermore, when the frequency characteristics of a machine tool are unknown, it is difficult to estimate the key vibrational frequencies to monitor.
2. The method proposed in this article, which uses the "variance of distance in force-vector locus" is effective and applicable for real-time supervision because the calculation only requires the sampling of data during one rotation of the cutting tool.
3. Since the variance of the distance varies with the cutting conditions, as well as with machining state, it is impossible to judge the machining state by only the absolute value of variance. Consequently the standard variance should be calculated using a physical model of the cutting process. The threshold variance is determined as a multiple of the standard variance.
4. In this article, the machining-state condition-sensing method is discussed in terms of the 2-dimensional force-vector locus. Generally, the method can be extended to n dimensions, for example, to a 3-dimensional force-vector locus, including the force along the machine tool axis, or to 6 dimensions, considering the 3 moment components. However, it is desirable to perform the necessary calculations in the sub-space (hyper-plane) in 6-dimensional space which characterizes the machining state, in order to reduce the calculation time which is required.

18.2.2.7 Adaptive Control

Evaluation method of cutting efficiency using the stability-lobe diagram. The spindle speed and the feed speed are proportionally controlled in the proposed system to maintain the feed-per-edge constant and maintain the constant surface roughness of the cut. Therefore, the equal metal-removal-rate curves are presented as hyperbolas where the multiplication of the depth of cut and the spindle speed is constant. In the stability-lobe diagram, at the upper right position the machining efficiency is high and, on the contrary, at the lower left position, it is low. However, it is rather difficult to find an optimum solution using the stability-lobe diagram because the boundary and equal metal-removal-rate conditions are both curves. This article proposes the following method to reduce the difficulty of finding the solution. Metal removal rate Q is represented as $Q = a\ b\ S_z Z\ n = a\ b\ V_f$, where a, b, S_z, n, Z and V_f represent depth of cut, width of cut, feed per edge, spindle speed, number of teeth, and feed speed, respectively. The relation $Q/b = a/t$ is obtained if t is determined as V_f^{-1}. Equal metal-removal rate Q can be presented as a straight line in the figure where t and a are selected as horizontal and vertical axes, respectively. The maximum metal-removal rate Q_{max} can be selected as $Q_{max} = b\ tan\theta_{max}$. Using this figure it is possible to evaluate both the stability and efficiency of the system.

Modification of the database
The stability-lobe diagram in the database does not always reflect the actual stability boundary because there are several unstable parameters including tool wear. Therefore, it is necessary to modify the diagram when the actual cutting state differs from the predicted cutting state. In the implemented system the boundary is shifted in the direction of depth of cut to satisfy the actual cutting state because the static cutting-force coefficient increases if the tool wear becomes significant [4].

Experimental results
Cutting experiments were performed to evaluate the developed system. The following rules were applied in the adaptive real-time controller: (i) The spindle speed and the feed speed were increased proportionally to increase efficiency while the cutting state remained stable. (ii) The spindle speed and the feed speed were decreased to avoid chattering, when it occurred.

In the experiment, the machining-condition determination task estimated depth of cut of 14 mm and spindle speed of 6,000 rpm as appropriate for the first cutting pass, towards the total depth of cut of 27 mm. Between 0 and 50 s, chatter vibration occurred because the tool was not sufficiently worn to provide frictional damping, and preventive control was executed. The spindle speed could be significantly increased at about 80 s and 130 s because light cutting conditions were encountered when the tool was changing the paths and the width of cut was locally decreased. From the experimental data, it was also clear that between 108 s and 125 s the cutting state was stable and cutting conditions remained almost constant. From this result, the system recognized that 2,940 rpm is the stable-unstable boundary for cutting with a depth of cut of 14 mm, which was obtained after 108 s. This result was used to modify the database.

Dynamic change of the database and the evaluation of the system

The 1st curve represents the initial estimated stability-lobe diagram determined from the database and derived from other similar experiments. For the current experiment, the stability-lobe diagram was modified in the 2nd curve to fit the actual conditions for a depth of cut of 14 mm and a spindle speed of 2,940 rpm, the limiting speed of the stable-unstable boundary. In the experiment, the machining-condition determination task was invoked again for the remaining 13 mm depth which remained for the overall, desired cut depth of 27 mm. The system determined that the machining condition should be 2,996 rpm for the spindle speed and 13 mm for the depth of cut. These machining conditions were used as initial conditions for the 2nd cut pass and adaptive control was performed again. In the experimental data, between 0 and 85 s, the initial cutting conditions were maintained almost constant even though the adaptive control was applied. The results showed that the previously determined cutting conditions were still appropriate. However, after 85 s, the system gradually evolved to a new stable-unstable boundary at 4,640 rpm in the spindle speed for a 13 mm depth of cut, as shown as the 3rd curve in the experimental data.

Summary

This article proposed and implemented a milling system where the tool path can be generated quasi-dynamically and the experimentally obtained stability-lobe diagram as a database can be modified during machining while using an adaptive controller to maintain the machining state stability using multi-axis-force information. Experimental results showed the effectiveness of the system. The proposed method in the article can be applied to all kinds of tools and workpieces.

18.2.2.8
Learning Control: Database Acquisition

Efficient database acquisition methods
Binary search method: Strategy I

1. **Overview of the method.** In this strategy the stability-lobe diagram is approximated by knowing the local maximum and minimum values [1]. They are called "diagram control points" in this article. These points are experimentally obtained by a binary search method in the spindle-speed and depth-of-cut coordinate frames. The stability-lobe diagram is obtained by a real-time machining condition-detection method [3], using adaptive-control capability for 2.5-axis cutting.
2. **Search algorithm.** The depth of cut is provided according to the following equation:

$$d_{ij} = \frac{2j-1}{2^{i-1}} l \tag{18.6}$$

$(i=1,2,...,D; j=1,2,...,2^{i-2})$

where i, j, D and l represent search depth, integer from 1 to 2^{i-2}, upper limit of search depth, and cutter length, respectively. Scanning area can be reduced because the shape of the stability lobe is convex to the lower direction. The lower part is always stable if the cutting state is stable at some depth of cut. On the contrary, the upper part is always unstable if the cutting state is unstable at the other depth of cut. To examine the stability at the depth of cut d_{ij}, the common area should be tested between the unstable area for just above ($d_{ij,\mathrm{upper}}$) and the stable area for just below ($d_{ij,\mathrm{lower}}$). The searching area $T(d_{ij})$ at the depth of cut d_{ij} can be written as follows:

$$T(d_{ij}) = \overline{A}(d_{ij,\mathrm{upper}}) \wedge A(d_{ij,\mathrm{lower}}) \tag{18.7}$$

where $A(d)$ and $\overline{A}(d)$ represent stable and unstable cutting areas in spindle speed direction at the depth of cut d, respectively. The stability-lobe diagram can be obtained by interpolating the "diagram control points" mentioned above and the stable-unstable limit at the designated depth of cut determined by the search tree. The search is completed when the search depth reaches D named "maximum search depth."

Heuristic method: Strategy II

1. **Overview of the method.** The aim of "Strategy I" was efficient acquisition of the diagram. However, to use the diagram as a database to find efficient cutting conditions, complete stability-lobe diagrams are not necessary. The information of the highest stable spindle speed at one depth of cut is crucial. Therefore, acquisition of the highest spindle speed in the stable area is required for a successful strategy for database construction.
2. **Search algorithm.** In "Strategy II" the diagram is divided into an appropriate size mesh. The spindle speed and the depth of cut are selected as discrete values such as r_i and d_j where $1 \le i \le M$ and $1 \le j \le N$. M and N present the total division number along each axis. It takes significant time if these numbers are large even though it is possible to obtain precise boundary information. The starting point for the search is selected as (r_M, d_1) to obtain the maximum spindle speed in the stable area. A cutting experiment is performed at the lattice points. The cutting condition is modified to move to the upper or left lattice point depending on the respective stable and unstable cutting state. The search process terminates when the spindle speed or the depth of cut achieves the allowed minimum spindle speed (r_1) or the allowed maximum depth of cut (d_N), respectively.

Experimental results and evaluation of the methods

Experiments were performed to evaluate the proposed methods. In "Strategy I" the maximum search depth was determined as 4. In the cases of "no strategy" and "Strategy II," the spindle interval and the interval of the depth of cut were determined as 500 rpm and 2.5 mm, respectively. Total cutting lengths and times were 440 mm (140s), 392 mm (123s), and 65 mm (18s) for "without strategy," "Strategy I," and "Strategy II," respectively. In "Strategy I" and "Strategy II," the diagrams were obtained efficiently compared with the case of "without strategy." In particular, the reduction ratio in "Strategy II" is outstanding. Tool wear might cause the difference of the shape of the obtained diagrams between "without strategy" and "Strategy I," even though other cutting conditions were the same. The main reason of the difference of "Strategy II" compared with the other diagrams was that only the highest stable spindle speed at one depth of cut was obtained. There is a possibility that the stable/unstable boundary might exist at the lower spindle speed. However, efficient cutting conditions can be determined using the obtained diagram without any problem as mentioned above.

18.2.2.9 Database Sharing

Database sharing method

In this section the database-sharing method among different machine tools is discussed. As the cutting stability of a system depends on mechanical properties of machine tools, such as the shape of the cutting tools, the geometrical shape of the workpieces, and the machine tool structures, the stability-lobe diagram is different for each machine tool. It is required, therefore, to obtain diagrams separately for various combinations of factors. However, this is inefficient. One solution to this problem is to share the database among multiple machine tools. According to [5], parameters which depend on tool wear are chip-thickness coefficient k_1 and penetration-rate coefficient K. k_1 transforms the diagram in the direction of depth of cut. Furthermore, it is known that K enlarges or shrinks the lobe in both the spindle speed and depth of cut directions. Therefore, the diagram after significant tool wear can be obtained by enlarging and shrinking the original one in the correct direction. Database-renewal method can be shared by considering that the magnification ratios represent the amount of the tool wear.

Database sharing experiment

Sharing of a database between two machine tools was tried in an experiment [1]. At first, the transition of the stability-lobe diagram caused by tool wear was measured on one machine tool. The tendency of lobe changes was used to estimate the other machine's lobe change after significant tool wear.

In the experiment, the magnification ratios were determined in the directions of spindle speed and depth of cut. The stability lobe was well predicted. In particular, in the lower spindle-speed range, the predicted curve fit exactly to the measured one.

Summary

This article discussed the implementation of machine tools with open-architecture CNCs for real-time machining-state monitoring, adaptive control and learning capability. The article also proposed rapid database acquisition methods for efficient cutting-condition determination. Furthermore, a database-sharing method among different machine tools to compensate for tool wear was described. The effectiveness of these methods was experimentally confirmed.

18.2.2.10
Active Thermal Deformation Compensation

Structure of the machining center

To measure the deformation of a vertical-type machining center, 4 deformation sensors are attached to the head, and 4 and 8 deformation sensors are attached to the upper and lower parts of the column, respectively, as shown in Fig. 18.11 [6]. To actively compensate for the tool position and attitude error, 16 thermal actuators which consist of electrical heaters and cooling jackets are attached at the lower part of the column. These thermal actuators do not reduce the rigidity of the system and it is easy to realize small angular adjustments because they do require bearings to produce translation and rotation. A combination of 2-layer deformation elements realizes translational motion in the *X*, *Y* and *Z* directions and rotational motion around *X*- and *Y*-axes. The position of the tool is measured by monitoring the motion of a cylindrical test bar (280mm in length and made of super-invar) using five eddy-current sensors.

The deformation sensor consists of a strain gauge-type detecting block and a connecting rod as shown in Fig. 18.12. The deformation between the left and right fixing points is concentrated in the deformation of three thin beams located in the center of the detector. Thus, small changes in the long distance between the two measuring points can be detected with high precision and stability. As the deformation sensor is made of super-invar metal with very low thermal expansion, only the thermal distortion of the machine itself is measured without influence of the thermal changes in the environment or in the machine structure.

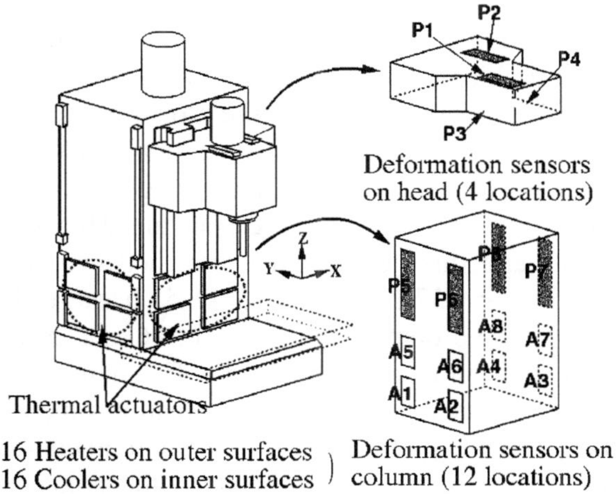

Fig. 18.11 Deformation measuring of a vertical type machining center.

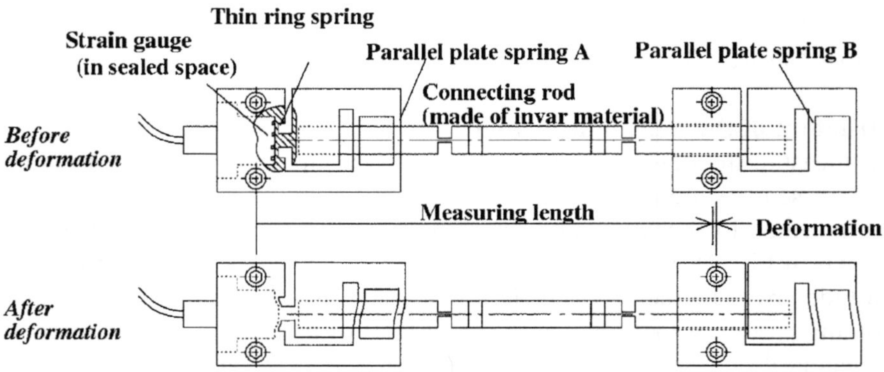

Fig. 18.12 Deformation sensor.

The performance of the thermal actuators is fast enough to actively compensate for the thermal deformation. To move the tip of the tool in 1 μm, it takes approximately 10 s.

Integration of neural networks and genetic algorithms for toll prediction

1. **Prediction of tool position.** In the system described in this article, the orientation error is compensated by the thermal actuators mentioned above, and the translational error is compensated using the origin-shift function of the numerical controller. Therefore, tool position prediction is needed. It is predicted by using output from deformation sensors because cutting chips and oil cause interference with the direct measurement of the tool position. The tool position is predicted by a multi-layer neural network. The inputs to the neural network are the 16 outputs from the deformation sensors, and the outputs are position and orientation of the tool. One of the problems with a neural network is that the solution may fall into a local minimum instead of the global minimum. This problem is avoided by using genetic algorithms mentioned later.
2. **Application of genetic algorithms.** To avoid local minima, genetic algorithms are used. The chromosomes of each gene in genetic algorithms reflect the connection weights and thresholds of a neural network. The length of a gene is determined as 357 (16×16 16×5 16 5) because the node numbers of input, hidden, and output layers are 16, 16, and 5, respectively, considering also the number of thresholds. The number of genes in one generation is determined as 25. As an evaluation function which represents the fitting capability of a gene to the environment, an inverse of the summation of difference squares between the predicted values and measured values as teaching signals is used. It is called fitness.

The next generation comprises descendants which are generated from the high fitness genes in the current generation. The selection probability is determined as 0.6 in the system. Furthermore, a homogeneous crossover method was applied. Mutation probability and the maximum generation were determined as 0.1 and 2,000, respectively. These values concerning genetic algorithms were determined by trial and error.

3. **Tool-position prediction experiments.** In the experiment, the tool position prediction capabilities between the neural network method and integration of neural network and genetic algorithms method were compared. These results were obtained by using the same sensor output data, a different set of data than that used for the training. The experimental results indicate that the integration of genetic algorithms and neural network improves the tool-position prediction accuracy.
4. **Active thermal deformation compensation.** To compensate for the tool posture error caused by the thermal deformation, neural networks as shown in Fig. 18.13 and Fig. 18.14 are used. Because it is difficult to measure the tool position and orientation during a cutting operation, the neural network has to have two modes: one is the "learning mode" and the other one is the "control mode" which is used during actual machining operation. In the control mode, control is executed using the weights and thresholds determined during learning-mode operation. The following equation is used as an evaluation function and the learning is executed to minimize the evaluation function:

$$E = \frac{1}{2}\sum_{c}(z_c - \hat{z}_c)^2 \tag{18.8}$$

where z_c, $\hat{z}_c$, and suffix c represent the current tool position and orientation, the tool position and orientation at the beginning of the experiment, and one of five measured positions of the test bar, respectively. The learning strategy is determined as follows:

$$\Delta w_{kj} = -\alpha \frac{\partial E}{\partial z_c}\frac{\partial z_c}{\partial y_k}\frac{\partial y_k}{\partial u_k}\frac{\partial u_k}{\partial w_{kj}} \tag{18.9}$$

where k, j, u_k, y_k and w_{kj} represent the output layer node, the hidden layer node, the summation of the outputs from the hidden layer multiplied by the weights, the outputs from output layer nodes, and the connection weights between the hidden layer and the output layer, respectively. α is a positive constant. The value of $\frac{\partial z_c}{\partial y_k}$ is determined in ad-

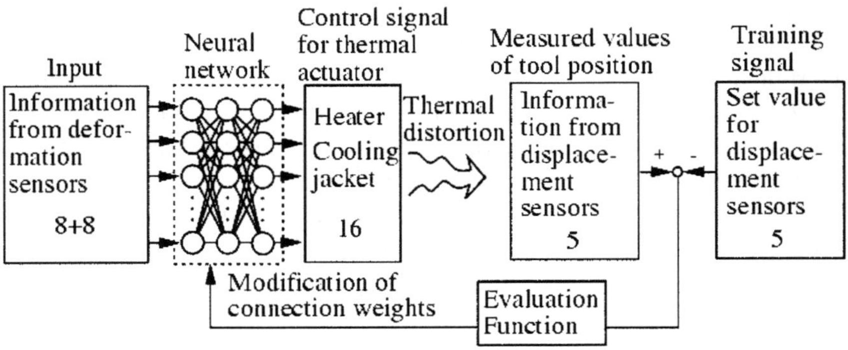

Fig. 18.13 Examples of neural networks for compensation of the tool posture error caused by the thermal deformaton.

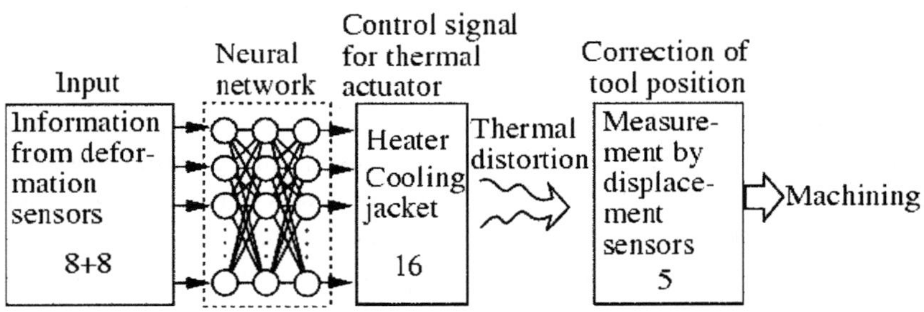

Fig. 18.14 Examples of neural networks for compensation of the tool posture error caused by the thermal deformaton.

vance by considering the relationship between the output of thermal actuators and the change of tool position and orientation. A 3-state extended sigmoid function is introduced because the output of the network may take on one of three values such as heating, cooling, and "no operation:"

$$f(u) = \frac{1}{2}\left\{\frac{1}{1+\exp(-(u-a)/c)} + \frac{1}{1+\exp(-(u+a)/c)}\right\} \tag{18.10}$$

where a and c are a parameter which controls the range of the middle value, and a parameter which controls the sensitivity of the network, respectively.

Figure 18.15 shows the experimental results of tool position and orientation using the methods described in this article. Figure 18.16 shows, on the contrary, the measured motion of the test bar without control. The spindle speed was 5,000 min^{-1}, the room temperature varied in a range of 14°C,

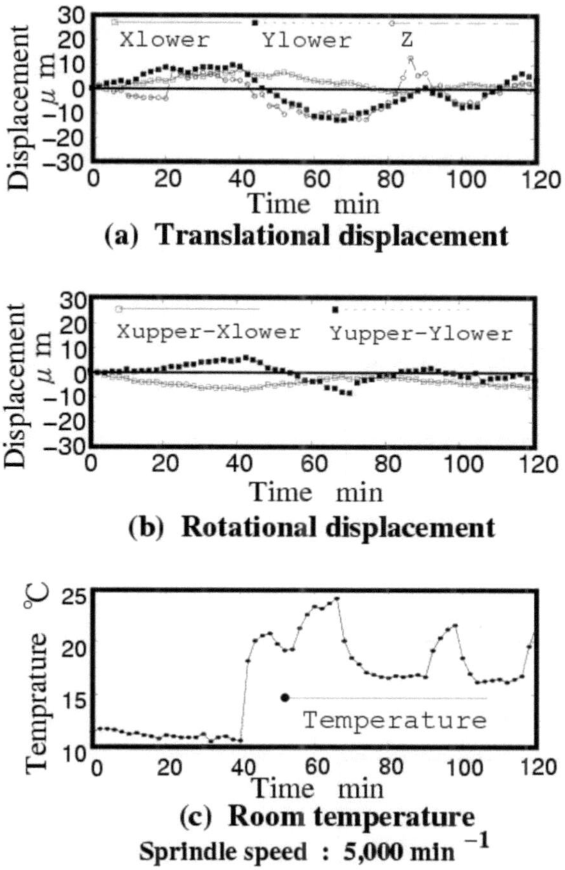

Fig. 18.15 Experimental results of tool position and orientation using the described methods

and the position and the attitude of the spindle were maintained within 12 μm and 21.4 μrad (6 μm/280 mm), respectively. The achieved error was less than 30% of the level of the uncontrolled machine.

Summary

A machining center that employs thermal actuators to actively compensate for thermal deformation has been developed which measures thermal distortion through an internal monitoring method using newly investigated deformation sensors. Using 3-layer neural networks integrated with genetic algorithms, although the room temperature varies in a range of 14 °C, the position and the orientation of the spindle is maintained within 12 μm and 21.4 μrad, respectively. The achieved error is less than 30% of the level of the uncontrolled machine.

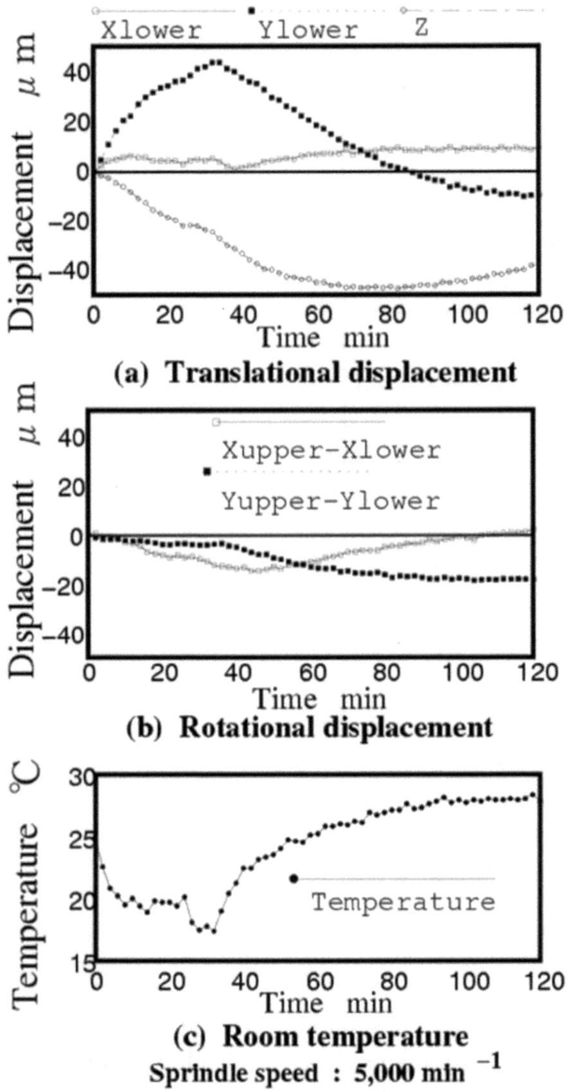

Fig. 18.16 Measured motion of the test bar without control.

18.2.2.11
Networked Manufacturing with Reality Sensation for Technology Transfer

The need for reality transmission to accomplish technology transfer
Global manufacturing is becoming increasingly feasible as information networks such as computer networks become widespread. This fact should drive basic changes in methods for technology transfer and education, as

well providing a means for the remote operation of machine tools. The manufacturing system developed in the article incorporates a reality transmission capability which provides an environment for the simultaneous transfer of know-how and experience without the restrictions imposed by place and time [7]. The technologies discussed in the article can be used to establish a remote operation and monitoring system with an advanced user interface that can be used for routine production systems in the future. The system developed by the authors consists of an open-architecture CNC controller; sensors to monitor the machining state; mechanical fuses to prevent overloads; apparatuses to present visual, auditory, force, and tactile information to the user interface; and an input device to operate a machine tool. The effectiveness of the proposed method was confirmed in experiments using the machine tools that were developed.

Conventional *tele-operation* enlarges the activity area of human beings by overcoming distance and scale restrictions. Conventional "technology transfer" means the transmission of technology from technologically advanced countries to developing countries over long distances. This process is typically very time-consuming. On the other hand, *learning* and *education* connote the transmission of existing knowledge and facts to the present and from the present to the future, respectively. Now, there is an increasing need for the global sharing of technologies and knowledge because of the increasing importance of the following items: environmental issues, food issues, teamwork in the medical area for tele-medicine and tele-surgery, disposal of industrial waste, dismantling of nuclear power plants, countermeasures against the accidents at nuclear power plants, and calamities, such as earthquakes, landslides, cave-ins, and oil tanker spills. Consequently, *simultaneous technology transfer* is becoming more and more crucial. Such transfer can be realized by remote manufacturing with reality transmission capability.

Operational environment transmission

Multi-modal operational environment transmission and time-delay

To realize the reality-transmission function, phenomena are detected by various kinds of sensors and the sensed information is transmitted to the operation room and presented to the human operator. Conventionally, the same type of information is presented as sensed. The visual information obtained by a TV camera, for example, is presented as visual information on the monitor TV. In the developed system, the sensed information is converted into different kinds of information, augmented and presented multi-modally. Information transmission links should be established to send information for all five human senses, with the appropriate control signals enabled.

Visual information presentation

1. **Predictive display of geometric information.** Generally, communication delay reduces the operability of bilateral systems because, for example, it destabilizes them. To overcome this problem, a predictive visual-information-display system has been realized for time-delay compensation. Because the shape of the workpiece changes dynamically during cutting, the geometric model must be revised in real-time. Therefore, the z-map-modeling method was introduced into the system.
2. **Real-time machiningsurface display using multi-axis force information.**The human operator may want to check the machined surface in real time to assure that machining is occurring under normal cutting conditions, but the sensitivity of the visual information collecting device (TV camera) is not high enough to display the surface of the workpiece. Moreover, the machined surface is covered with machining chips so that the operator can not see the surface precisely even if the sensitivity of the TV camera were high enough.

In the developed system, the visual information was derived through the transformation of measured-force information. In the visualization, the mapped cutting forces at each vertical line, which is perpendicular to the tool feed direction, are normalized. In this normalization, the force data are converted linearly in a scale between 0.0 and 1.0, and the gray-scaled color corresp/nding to each normalized value is assigned to each of the force-data points. The normalized force value of 0.0 corresponds to white, and that of 1.0 to black and the rest to gray in 8-bit depth. The interval between lines along the tool-feed direction can be calculated from the tool-feed speed. The force-to-visual-information-transformation technique produces augmented visual information, while the measured surface roughness transformed into a bit-map figure. Comparing these two pieces of augmented visual information, it is clear that the texture which is transformed and augmented from the cutting force is quite similar to that of the machined surface of the workpiece.

Auditory information presentation

In machining, generally, auditory information is very important in allowing the operator to recognize the cutting state. The frequency of the sound is determined as $\frac{rn}{60}$ using the spindle speed r and the number of teeth on the cutting tool n. If chatter is expected, the sound is generated using the preliminary acquired cutting force during chatter vibration. The cutting state, whether normal or chatter, is predicted using an experimentally obtained stability-lobe diagram.

Force-information presentation

1. **Force-smoothing method.** It is not appropriate to present the cutting force directly to the operator because the cutting force varies during one tool rotation even if feed speed remains constant. The authors propose a force-smoothing technique to display the varying machining force as a smooth force to an operator. In this technique, the center of gravity at the force-vector locus is selected as the representative cutting force during one rotation of the machine tool. The average of the representative force is displayed as the reflecting force along the direction of tool feed to enhance the operability of the system, although the actual resultant force directions differ from the feed direction.
2. **Predictive force presentation.** In the case of remote machining, substantial time delay is introduced by the transmission of information from the machining site to the operation room. This time delay may destabilize the system. If the remote working environment can be identified at the operation room, the identified information may be used to control the local multi-axis joystick. This control strategy causes almost no time delay because the control loop is closed in the controller located at the operation room. Figure 18.17 shows the control strategy of the system. The controller in the operation room has a map describing the relationship between the feed speed and the cutting force calculated from the physical model of cutting. The force applied by an operator to and the velocity information v^*, and last, the desired velocity of the joystick u^* is determined by the difference between the applied force F_j and the calculated virtual cutting force F_c^* as follows:

   the joystick is measured by a multi-axis force sensor. Velocity control is adopted for the control of the joystick because the viscous force is dominant when we consider the machining tool as a mechanical impedance. At first, the velocity of the joystick u is determined based on the measured force information F_j from an operator. Next, the virtual machining force F_c^* is calculated using the physical model of machining

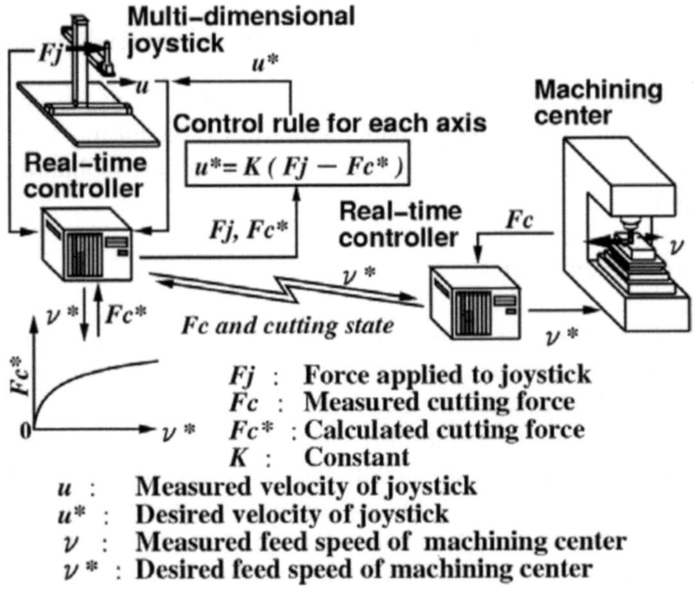

Fig. 18.17 Control strategy of the system.

$$u^* = K(F_j - F_c^*) \tag{18.11}$$

where K is constant. The feed speed of the cutting tool v^* is calculated to be proportional to that of the joystick u. It is transmitted from the operation-room controller to the machine-site controller through the communication system.

Tactile information presentation

Tactile-information presentation of machining state

In this article, the machining state such as chatter vibration, is presented as a vibration. To present the machining state, the following value is calculated in real-time and is sent from the machine site to the operation room:

$$\frac{V(|F - G|)}{|G|^2} \tag{18.12}$$

where V, $\boldsymbol{F}$, and $\boldsymbol{G}$ represent the variance during one tool rotation, cutting force vector, and the center of the gravity of cutting force during one tool rotation, respectively. The value $V(|\boldsymbol{F} - \boldsymbol{G}|)$ is used to judge the machining state in real-time [3]. The value becomes large when chatter vibration occurs. However, the value changes depending on machining conditions such as feed per edge S_z, even if the machining state is normal. Using the physical model of cutting, $V(|\boldsymbol{F} - \boldsymbol{G}|)$ and $|\boldsymbol{G}|$ can be calculated as follows, respectively:

$$V(|F - G|) = C_v S_z^{2(1-\varepsilon)} \tag{18.13}$$

and

$$|G| = C_g S_z^{1-\varepsilon} \tag{18.14}$$

where ε is an exponential coefficient depending on the shape of a cutting tool, tool material, and workpiece material. C_v and C_g are proportional constants depending on depth of cut and width of cut as well as the cutting conditions mentioned above. An eccentric weight and a motor were installed at the grip of the 2+1 dimensional joysticks, and the rotational speed of the motor was controlled to be proportional to the value described in equation (12). The operator feels vibration when chatter vibration occurs.

Summary
A networked remote manufacturing system with reality-transmission capability has been developed. It consists of an open-architecture CNC controller; sensors to monitor the machining state; mechanical fuses to prevent overloads; apparatuses to prevent visual, auditory, force, and tactile information to the user interface; and an input device to operate a machine tool. The developed system provides an environment for the simultaneous trans- fer of know-how and experience without the restrictions imposed by place and time.

18.2.3 Development of an Intelligent Face Grinding Machine

M. Nakao, Y. Hatamura

18.2.3.1 Introduction

Information-processing devices handle thin substrates, such as silicon wafers, glass lithography masks, ceramic circuit boards, and glass display boards. These substrates are 8 inches or larger in diameter with ultra-flatness of some 0.1 μm per 200 × 200 mm.

Current processes for finishing the substrates are lapping and polishing. These processes suffer from a low material-removal rate, and large machine footprints, e.g., a 3 m lapping plate is used in a 2-h batch process to average out variation in the removal; an alternative is individual grinding. Various factors interfere with conventional grinding of such substrates: the grinding force causes deformation at the rotary bearing; irregular thermal expansion bends the column; the table tilts along the linear guide, and consequentially, conventional grinding produces convex substrate surfaces with flatness within about 10 m. Some studies have shown that more rigid bearings, a thicker column, a tighter room-temperature control, and hydrostatic linear guides can reduce tool-positioning error [8, 9]. Such passive methods produce certain effects; however, even the most expensive one cannot realize a zero positioning error. In contrast, an active method can realize a zero positioning error. For example, equipped with sensors to detect structural deformation in the column and compensating actuators, even if the machine deforms, the computer, which knows all the "causes and effects" of environmental variables to tool positioning, can precisely control the latter by measuring the former [10].

To accomplish ultra-flatness with active control, the authors designed and fabricated an "intelligent face-grinding machine". The machine was a vertical-spindle face-grinding machine equipped with knowledge-based control with sensors, controller, and actuators actively controlling the tool position to eliminate deviation. We then evaluated our system.

18.2.3.2
Designing the Intelligent Face-Grinding Machine

To design the intelligent face-grinding machine, we sequentially set the following factors: requirements, functions, elementary tasks, individual mechanisms, structural components, and finally the total system [11]. They were: i) Requirement of "flatness is 0.1 mm per 200 × 200 mm."; ii) Functions for analyzing if requirement is met: "force-induced-deformation compensation," "thermal-deformation compensation," "bearing-fluctuation compensation," and "zero-deformation fixture"; iii) Elementary tasks decomposed from their functions such as displacement sensing or fine actuation; iv) Individual mechanisms proposed for each task, such as strain gauges or piezo elements; v) Structural components developed for each mechanism; vi) Synthesis of the total system using all components.

Figure 18.18 shows a schematic view of the total system. The column is made of cast iron and weighs 83 kN. We set the coordinates as shown in the figure: X – longitudinal feeding direction, Y – transverse, and Z – vertical.

Higher rigidity and thermal insensitivity of the structure passively reduces tool-positioning error. Further, the firm structure helps activate controllability with less variables to detect and a narrower control range. For example, the structure deforms symmetrically with its 4-leg columns, and temperature distribution is uniform with the thermally insensitive column surface to make the "thermal-deformation compensation" easier (Sect. 18.2.4.2).

Figure 18.19 shows a conceptual sketch of individual mechanisms. Section 18.2.4.3 describes the main mechanisms in detail: double actuation mechanism for "spindle posture control" and "table posture control," and two sensors for detection of "thermal deformation" and "wheel tilt."

18.2.3.3
Developing Structural Components

This section explains major components we developed. Other components and the control algorithms are explained in our previous work [10–13].

The two main actuation mechanisms use multi-layered piezo elements for fine positioning. Piezo elements 20 mm long generate a 10 m displacement and up to 3.5 kN force under 100 V. Strain gauges attached on thin plates (Figure 18.19) feed back the displacement signal. The feedback control makes the piezo element displacement independent of voltage history. Further, the control stiffens the machine infinitely; the piezo elements expand over the same volume as the force-induced shrinking of other components [14].

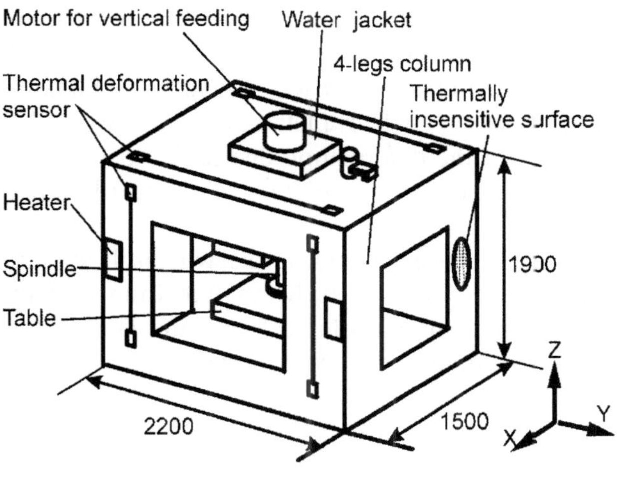

Fig. 18.18 Schematic view of the intelligent face grinding machine.

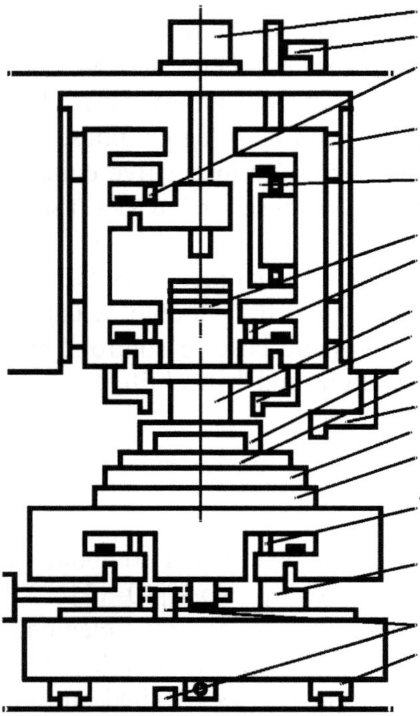

Fig. 18.19 Aconceptual sketch of the intelligent face grinding machine.

Using four pieces of 60-mm-long piezo elements, the spindle posture-control mechanism controls the depth of cut to a 30 nm resolution within the 29 m work range and the wheel tilt to 0.3 rad within 240 rad. Similarly with four pieces of 20 mm-long piezo elements, the table posture-control mechanism controls the Z height, pitch, and roll, respectively to a 10 nm resolution within the 12 m work range, 0.1 rad within 62 rad, and 0.1 rad within 44 rad. The thermal-deformation sensor made of Invar material does not thermally expand like the machine made of metal. Strain gauges on the sensor, thus, provide a measure of the deformation due to thermal expansion. The sensor detects the expansion to a resolution of about 0.1 m per 1,000 mm.

Four capacitance-displacement sensors compose the wheel-tilt sensor. The gap between the wheel and the sensor is always free from dust or moisture because the air flow from the spindle seals off air. The sensor measures Z-height to a 10 nm resolution and tilt to 0.03 rad.

18.2.4
Evaluating the Intelligent Face-Grinding Machine

18.2.4.1
Compensating for Force-Induced Deformation

For face grinding to produce ultra-flat surfaces using a straight-cup wheel, it is most important to keep the wheel surface normal to the spindle axis. Otherwise, a tilted wheel generates a curved surface on the workpiece.

Figure 18.20 shows normal grinding-force history under the conditions in the figure; the force history indicates that the upward-pushing force on the wheel initially increases as the wheel engages the workpiece, and decreases toward the end; Figure 18.21 shows that the force pushed the wheel upward by 3 m and the moment turned the front side up by 12 rad. Figure 18.22 shows the surface profiles after grinding: convex curvature of 2.0 m in the longitudinal feeding (X) direction and 2.5 m in the transverse feeding (Y) direction. The profiles were measured with the work fixed on the table. High rigidity of the machine, 60 N/m in Z direction and 2 Nm/rad in Y rotation, improved surface flatness from about 10 m to 2.5 m. Passive mechanisms alone cannot accomplish such ultra-flatness of 0.1 m.

Next we kept the wheel flat using the wheel-tilt sensor and the spindle posture-control mechanism. Figure 18.23 shows the wheel posture under control; the wheel stayed within ±0.3 m of the initial Z position and ±2 rad rotationally. Figure 18.24 shows surface profiles after grinding with this control: convex curvature of 0.8 m in the longitudinal feeding (X) direction

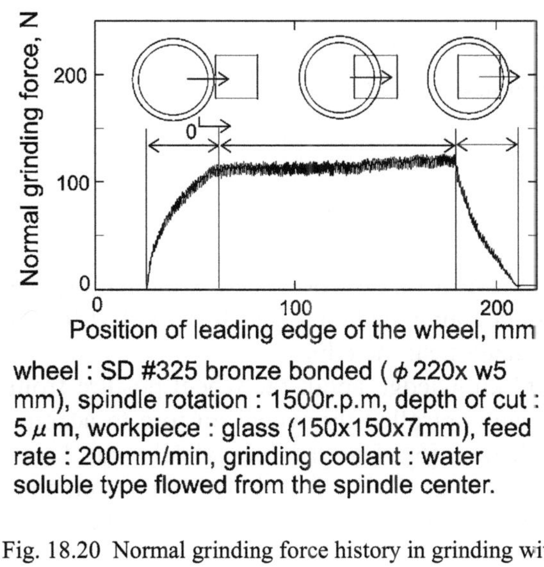

Fig. 18.20 Normal grinding force history in grinding without any posture control.

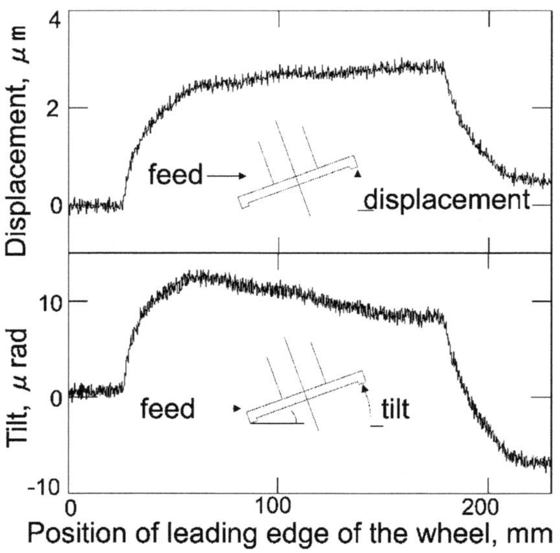

Fig. 18.21 Displacement and tilt of the wheel in grinding without any control.

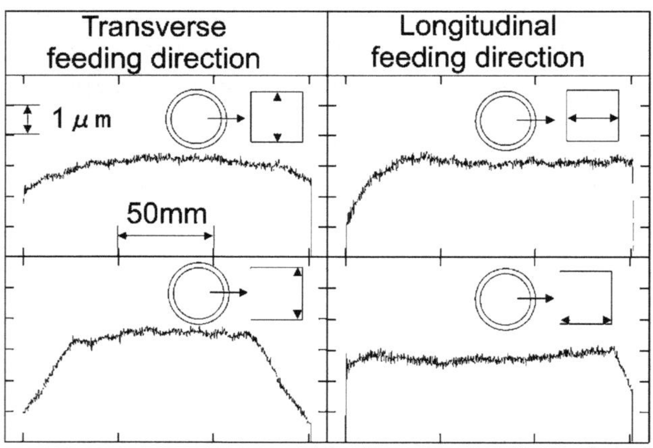

Fig. 18.22 Profile of the ground workpiece in grinding without any control.

and 0.6 m in the transverse feeding (Y) direction. Our grinding experiments suggest that we can produce an 8-inch size substrate with a 0.5 m flatness except in the 5mm perimeter near the edge.

Using the "cause and effect" knowledge about the wheel posture from Figs. 18.20, 18.21, and 18.22, we controlled the wheel posture to accomplish 0.8 m on a 6-inch size substrate [5]. The grinding wheel wear, however, always progresses: the condition during learning differs from that during controlling. For this reason, we controlled the wheel posture in a real-time manner.

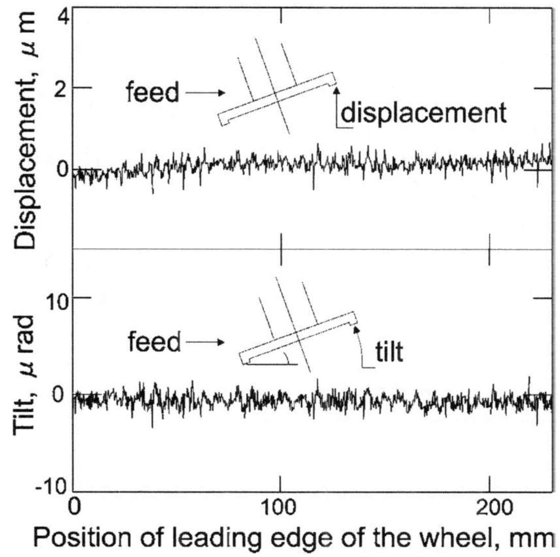

Fig. 18.23 Displacement and tilt of the wheel in grinding with the wheel posture control

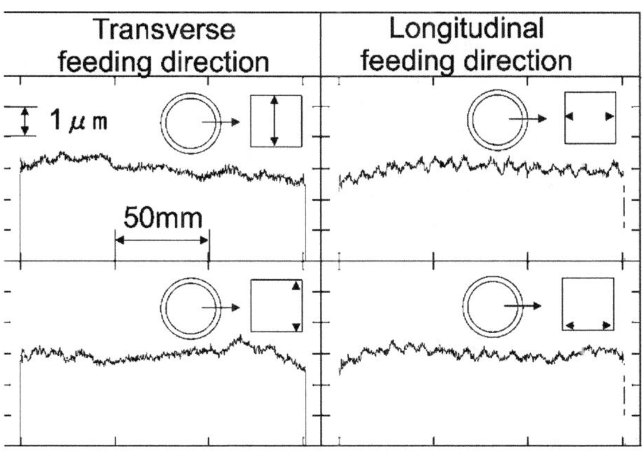

Fig. 18.24 Profile of the ground workpiece in grinding with the wheel posture control.

The figures show a vibration of 7.5 mm wavelength (or 0.5 Hz) with 0.3 m p-p amplitude in the Y profile (Fig. 18.24). The linear ball bearings were responsible for these vibrations (Sect. 18.2.4.3).

18.2.4.2 Compensating for Thermal Deformation

For producing ultra-flat surfaces, we need to keep the wheel tilt flat despite the column deformation that turns it.

Figure 18.25 shows the relations among room-temperature variations, thermal deformation of the legs, and the wheel displacement and tilt. When the thermally insensitive surface is not applied, an 18° change in room temperature causes the front leg #1 to expand 70 m longer than the back one #3. This causes the wheel to lift 14 m and tilt backwards by 16 rad. This posture change causes the variation shown in Fig. 18.21.

In addition to room temperature, the motor-coolant-flow variation added 2 rad and the spindle rotation added 4 rad. From Fig. 18.25, we observe that the wavelength of 1 h is much longer than the 1 min grinding

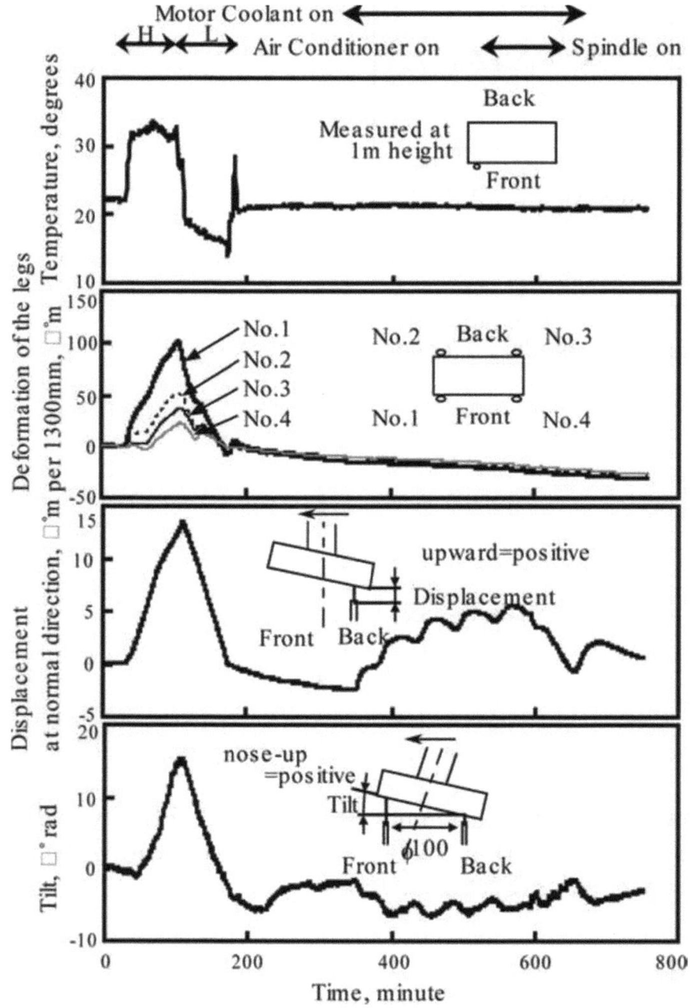

Fig. 18.25 Relation among room temperature variation, thermaldeformation ot the legs, and wheeldisplacement and tilt.

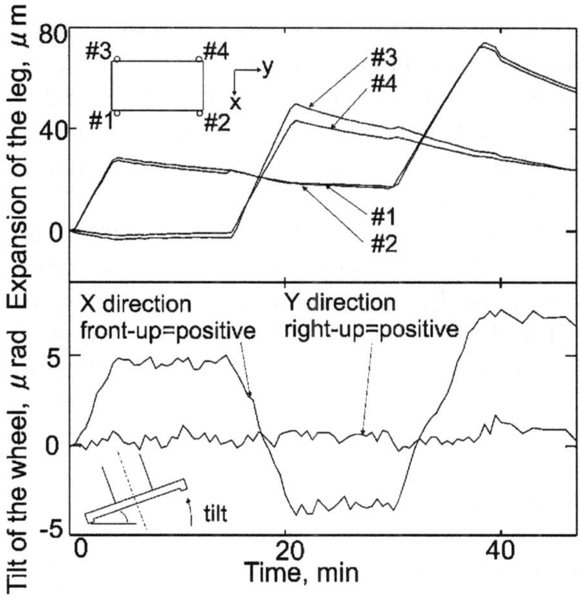

Fig. 18.26 Tilt of the wheel induced by expansion of the legs using the heaters.

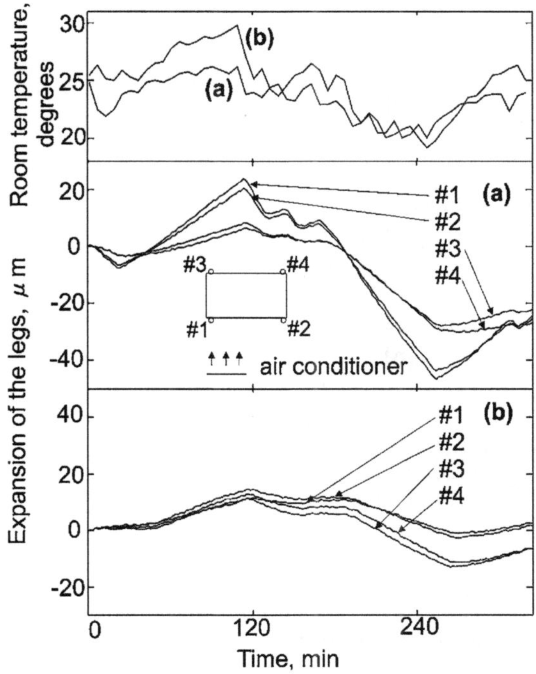

Fig. 18.27 The thermal expansion of the legs (a) without / (b) with thermaily insensitive surface.

time; however, thermal deformation gradually causes a negative effect; only a 1 rad tilt of the 220 mm wheel can degrade the flatness by 0.1 m.

Heaters on the column legs (Fig. 18.18), in addition to the spindle-attitude control mechanism, effectively countered this deformation [3]. Figure 18.26 shows that the wheel tilted by 5 rad when the heaters extend the two front (or back) side legs by 30 m.

We, however, found another passive method more cost-effective than this active method. Figure 18.27 shows the difference in thermal expansions of the legs with and without the thermally insensitive surface made of a 30-mm thick foam plastic. The figure shows only 15–30% of the original elongation. The plastic allows us to neglect thermal deformation of the column.

18.2.4.3
Compensating for Bearing Fluctuation

The linear ball bearings of the table also affect surface flatness. Figure 18.28 shows the Z-height deviation with and without compensation for five measurements. The curves without compensation show large repeated waves and spikes in smaller periods. The spikes show little repeatability

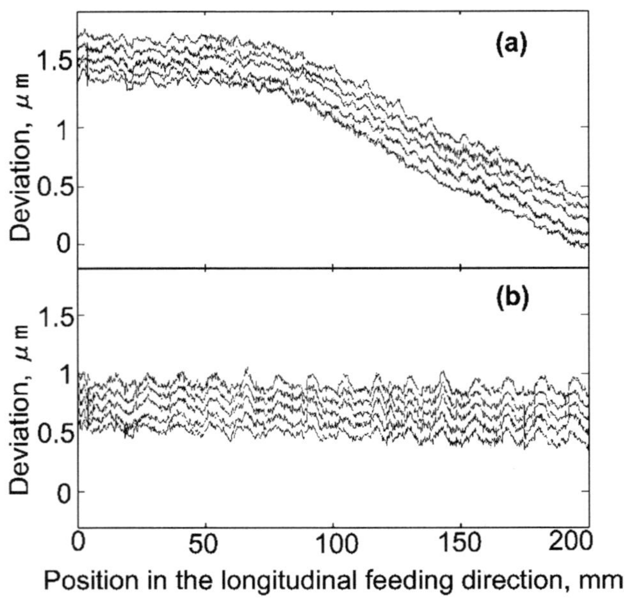

Fig. 18.28 Z-height deviation with and without compensation for five measurements.

and their intervals were twice the ball diameter. Thus, we conclude that these spikes were caused by movement of balls when they travel from the loose return groove to the preloaded slider groove. Figure 18.28 shows the worst case when the balls made some spikes while learning the original deviation but none when controlling for its compensation.

We first used a finite-feed range slider to solve the problem. With its balls always preloaded, the finite-feed range slider eliminated the vibration; however, it required longer rails and sliders than the infinite one.

Next we measured, in advance, vibration on a flat glass bar set next to the workpiece to compensate for the variation in the Z direction. This method allowed us to cancel the vibration during grinding, producing a surface with Z-height deviation within $\pm$0.05 m over an area of 250 × 250 mm.

18.2.4.4
Zero-Deformation Fixture

A flat vacuum chuck firmly holds a wavy substrate to produce a perfectly flat surface; however, upon releasing the chuck, the surface recovers the original wavy shape by its spring back. A common practice in the industry is to use a wax chuck, which still does not perform to our full satisfaction. Take, for example, a 150 × 150 × 7 mm quartz substrate. Chucking it with wax on an Invar table without any preload, we ground it, and once released, the spring back produced a 4-m change in the flatness. Although 3 m of it was due to residual ground stress that we could cancel by grinding the other side, the remaining 1 m was still due to imperfection in the fixture. We have tried a variety of different chucks: wax as mentioned above, ice by solidifying water between the table and the workpiece, and a pin array composed of a number of pins with their heights controlled individually to conform to the workpiece topography. All these trials still resulted in about a 1-m flatness change with the chucking.

We obtained the best results with a "plaster chuck." This chuck has a number of plaster bumps, a seal along the edges, and vacuum pressure. The type of plaster used, common for dental purposes, changes its volume by 0.02% after solidification. With this plaster chuck, we reduced flatness change with the above quartz substrate down to 0.2-0.5 m as Fig. 18.29 shows.

18.2.4.5
Total Process Evaluation

We evaluated our process by grinding the 150 × 150 × 7 mm quartz glass substrate with 220 mm (Fig. 18.18). The process follows: (a) Measured the

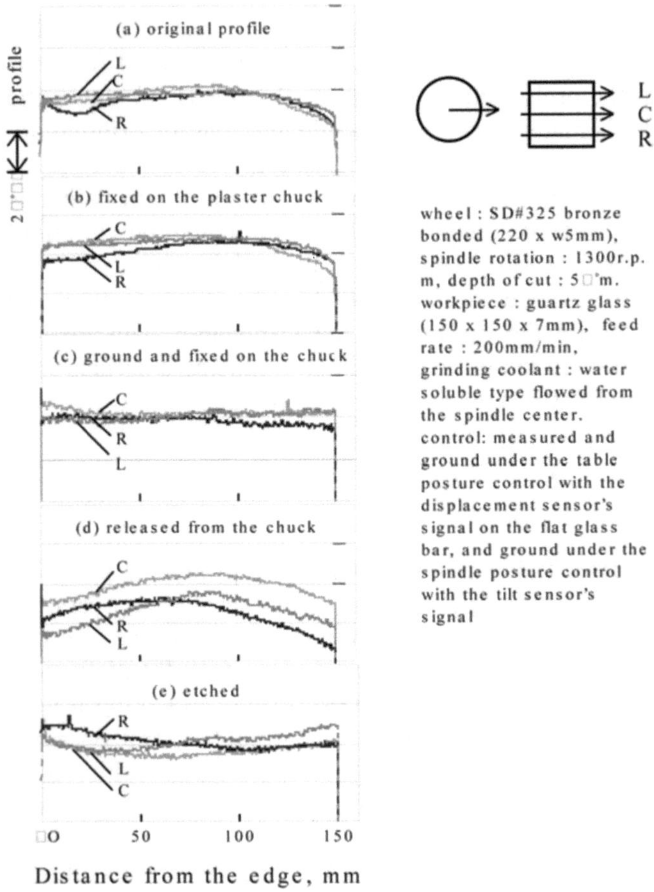

Fig.18.29 Profiles of the substrate on the grinding process.

original surface of a polished substrate without any loads, (b) fixed the substrate on the plaster chuck and measured surface, (c) ground the substrate and measured its surface on the chuck, (d) released the substrate from the plaster chuck and measured surface without any loads, (e) etched the substrate by hydrofluoric acid and measured its surface without any loads.

Signals from the tilt sensor were input to the spindle-posture real-time control mechanism, and signals from the displacement sensor on the flat glass bar to the table-posture real-time control mechanism.

Figure 18.29 shows the results. From (a) and (b), the plaster chuck deformed the work by about 0.5 m. From (c), the grinding flattened it to about 0.5 m. From (d), residual stress deformed it by about 2.5 m. From (e), the etching still left the stress; the 1.5 m twist remained.

Annealing the substrate on a flat surface will enhance the flatness by about 0.5 m because the process will release residual stress perfectly. However, the original requirement of "flatness of 0.1 m per 200 × 200 mm" is not yet realized. To further improve flatness, we need more knowledge about grinding-rate change of the wheel and residual-stress variation in the workpiece.

18.2.5
Conclusion

To produce a 0.1 m flatness on substrates for information-processing devces, we designed, fabricated, and evaluated an "intelligent face-grinding machine." The machine executes knowledge-based control using sensors, a controller, and actuators. Intelligent control by our machine improved the flatness of ground quartz glass substrate of 150 × 150 × 7 mm up to 0.5 m. Further knowledge about grinding-rate change and residual-stress variation will further improve the flatness from 0.5 m to 0.1 m.

We thank THK Co., Ltd., for fabricating the machine.

18.3
Modern Concepts for Machine Tools

F. Rehsteiner

In this chapter, an arbitrary selection of three novel developments in the design of machine tools is presented. All of them concern the machine tool properly speaking, i.e., neither software and communication aspects nor networked systems like flexible manufacturing systems are addressed. These latter fields, demanding as they may be, at least in their function of presenting a "background support" to the actual production work, have reached a high level of maturity and therefore need not be discussed here.

All of the three selected topics: Non-Cartesian machine-tool kinematics, doing away with idle time between different processes, and improving the dynamics of large machine tools, cannot be considered as "mature" by any means at the time of this writing. This holds true for their technology as well as their acceptance by the users. The intention here is to give the reader an idea of "where could the voyage go," and to show him that quantum leaps forward are still possible.

18.3.1 Machine Tools with Non-Rectangular-Axis Systems

18.3.1.1 History

One of the most important quality aspects of machine tools for cutting operations is their high accuracy. The shape of the workpieces produced by them is determined uniquely by the shape of the cutting tool combined with its motion. The tool motion is carried out by the machine tool. Thus the accuracy of the machine tool is transferred to the workpiece exactly one-to-one and is decisive for its accuracy.

Many workpiece geometries require a more complicated tool motion than simply moving along a straight axis; often several simultaneous "axis motions" are required to create even simple geometries. Therefore most machine tools comprise more than one axis of motion, and their combined accuracy is crucial.

In rectangular-coordinate systems the normal projections of a point onto the coordinate axes are independent from each other; see Fig. 18.30a. Therefore, in Cartesian machine tools each axis and their mutual positions can be calibrated independently. This is not possible in non-rectangular systems; see Fig. 18.30b.

Therefore, in order to achieve the utmost in accuracy, until recently almost all machine tools relied on a rectangular set of axes of motion; see Fig. 18.31a and b. Robots working with non-rectangular axis systems like, e.g., the bending arm robots shown in Fig. 18.31c, are used almost exclusively for pick-and-place tasks where a high positioning accuracy is less important than their geometric versatility.

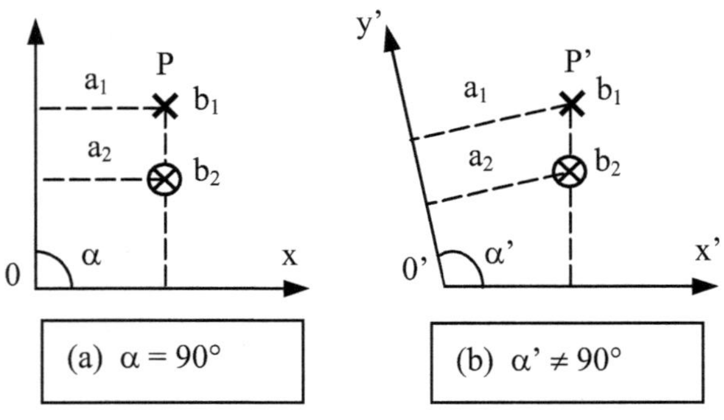

Fig. 18.30 If, in a rectangular coordinate system (a) x,y ($\alpha = 90°$) the distance b of a point P to a reference axis, say, x, is changed from b_1 to b_2 this does not affect its distance form the second.

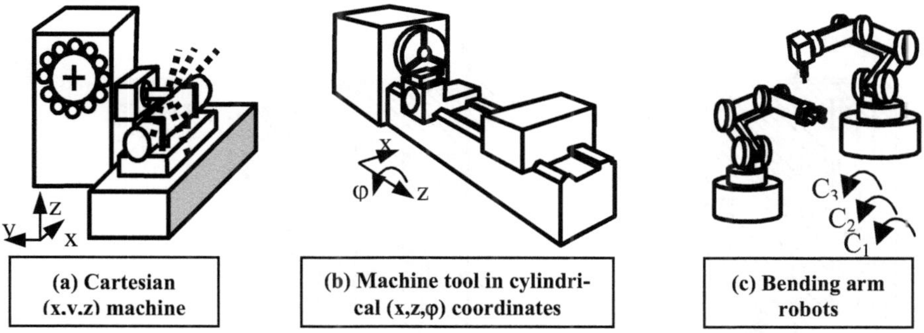

Fig. 18.31 Classical machine tool geometries. Most actual manufacturing machines like milling and boring machines, machining centers, and lathes work in rectangular coordinate systems (Figures (a), (b)). The bending arm robots, employing purely angular coordinates, have been used mostly for pick-and-place and for assembly tasks where high accuracy is not important. Further on, all of these machines are specialized towards one single process.

Most actual manufacturing machines like milling and boring machines, machining centers, and lathes work in rectangular coordinate systems (a, b). The bending arm robots, employing purely angular coordinates, have been used mostly for pick-and-place and for assembly tasks where high accuracy is not important. Further on, all of these machines are specialized towards one single process.

This restriction of machine tools to a rectangular-geometry base has lost its importance dramatically with the advent of powerful computers which perform even quite complicated coordinate transformations in real-time. Kinematic arrangements similar to the ones shown in Fig. 18.31c in which all of a sudden, six degrees of freedom are no more difficult to handle than one, have become serious candidates for machine tools, aside from problems like stiffness, calibration, thermal sensitivity, etc. This opened the doors to new paradigms about acceptable laws of motion for machine tools as discussed in the following section.

18.3.1.2
To Position a Rigid Body in Space

For machine tools, usually the kinematic task is one of positioning two partners, i.e., a tool tip (more precisely "the tool point creating the workpiece surface") and a workpiece relative to each other. This means automatically that both partners have to be positioned accurately in absolute space, as well.

In the logic of investigating non-rectangular machine-tool designs – although history worked differently – 2-DoF (Degrees of Freedom) arrange-

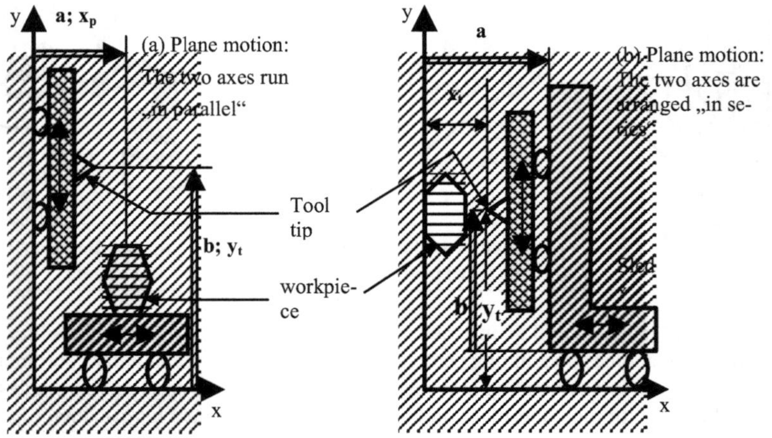

Fig. 18.32 Two classical ways of realizing 2 DoF movements. The arrangement of the axes of motion "in parallel", Figure (a), yields better dynamics and higher accuracy but is less suited for automation than the "serial" arrangement (b). This was a strong motivation for trying to build 3 DoF and even 6 DoF "Parallel Kinematic Machines" (PKM's) as exemplified by the Figures as exemplified by the figures 18.33 to 18.36.

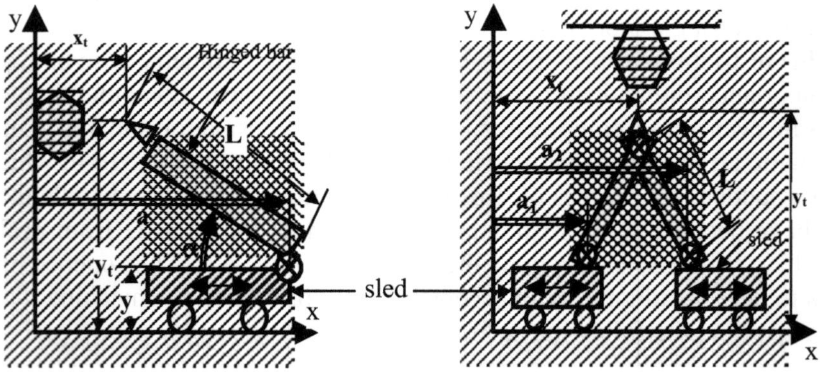

(a) "Hinged vertical axis" (b) "Scissor" arrangement

Fig. 18.33 Two examples of 2D positioning devices with hybrid kinematics. Both involve at least one angular motion: In the "hinged vertical axis" (a) the displacement in y is achieved by a rotation of the "hinged bar". In the "scissor" arrangement (b) both the x and y displacements of the tool tip result from a combination of the collinear movements of the two sleds.

ments are a good starting point. In a very simple arrangement, the two partners move along axes at right angles to each other, both fixed directly to the machine structure, as shown in Fig. 18.32a. This allows them to optimize both axes independently with respect to dynamics, accuracy, etc., thus yielding a very good performance. Since the workpiece is not stationary, however, it is not so well suited for automation. This has been corrected in the design shown in Fig. 18.32b. But here the dynamics are hampered by the fact that sled x is loaded not only by its own mass but also by the complete y axis. It is therefore not so well suited for very fast movements.

Recently the strive for ever higher dynamics has led to designs in which some of the linear base movements have been replaced by angular ones: In Fig. 18.33a the inertia of the hinged bar, the angular motion of which creates the y movement of the tool tip fixed to its "far end," is only 1/3 of its mass felt in linear motion. This pays off directly in improved dynamics.

Tool-tip position in y: y_t is controlled by the (power-driven) angle α: $y_t = y_0 + L.\sin\,\alpha$. Since this angular motion also influences the x position of the tool: $x_t = a - L.\cos\,\alpha$ (a being the powered x position of the sled), controlling its position in 2-D space is considerably more complicated than in simple linear motion.

An example of the large family of "scissor"-type kinematics is shown in Fig. 18.33b. Here, as in Fig. 18.33a, the y position yt is controlled by the angular position of the two bars. These in turn follow from the independent but collinear movement of the two sleds: $y_t^2 = L^2 - [(a_2 - a_1)/2]^2$ which in x leads to the simple relation $x_t = [(a_1 + a_2)/2$. This results in quite a simple design but in a limited workspace in y.

In the kinematic arrangement of Fig. 18.33b, both powered components, i.e., both sleds, only carry a hinged bar plus part of the tool-carrying element. These are passive elements which are much lighter than powered elements like the vertical sled in Fig. 18.32b. Therefore they are well suited for high dynamics.

Such designs in which nowhere two or more power axes are arranged in series, i.e., on top of each other (as they are in Figs. 18.32b and 18.33a) are called *Parallel Kinematic Machines* (PKM).

Because of this potential for very high dynamics, designers have been looking for "parallel" arrangements able to control up to 6 DoF. Rigid body mechanics are a good starting point.

One of their base rules says that the position of a rigid body is determined uniquely by 6 properly chosen constraints. Figure 18.34 shows this in a plausible, though not scientifically flawless, way.

This "six-point theory" holds true in a surprising variety of choices of the six "coordinates" which do not need to be Cartesian at all. Two examples of non-Cartesian 6-DoF arrangements are shown in Fig. 18.35.

Often one would prefer to work with 3 DoF only; this would particularly make NC programming much easier. Figure 18.36 shows a "Tripod" (three-legged device) in which the superfluous" rotary degrees of freedom of the platform are suppressed explicitly by the torsion bar. It features a large workspace, is reasonably fast, but lateral (x, y) stiffness is rather

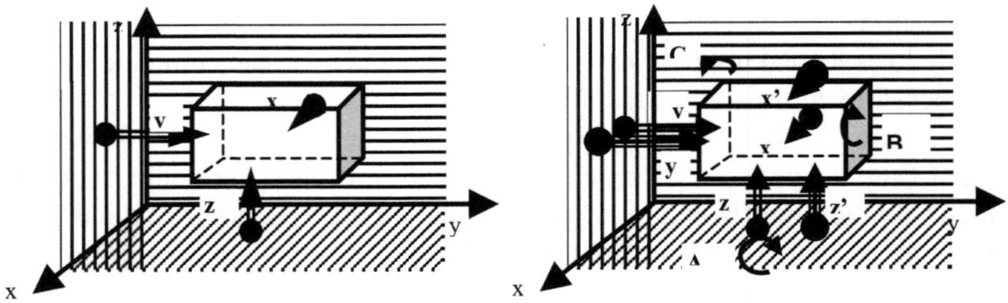

(a) Position only (b) Position and orientation

Fig. 18.34 The *position* of a rigid body in a Cartesian coordinate system is determined uniquely by 3 properly chosen points (x.y,z) on its surface, see Figure 3(a) (the problem of singular arrangements caused by improper points is left aside here). Three more points (x', y', z') are needed to fix also its *angular orientation* (A,B,C): (x,x') ⇒ B; (y,y') ⇒ C; (z,z') ⇒ A.

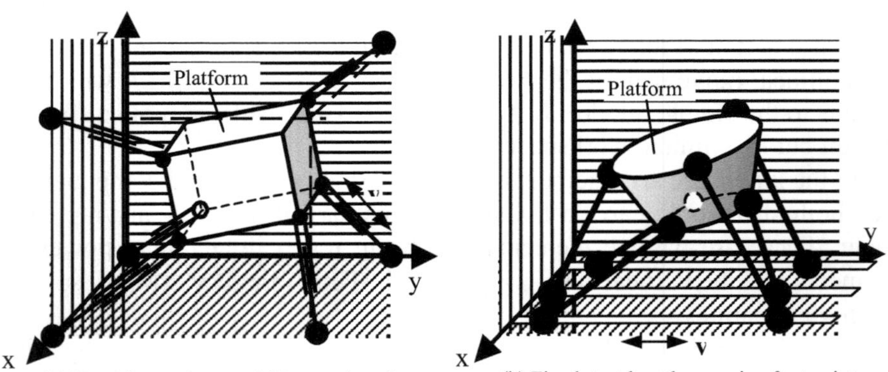

(a) Fixed foot points, variable strut lengths (b) Fixed strut lengths, moving foot points

Fig. 18.35 Two arrangements of "Hexapod" kinematics. The Greek word Hexapod means "six legged". In both of them the range of the 6 Degrees of Freedom: x,y,z,A,B,C (see Fig. 5 (b)) is limited, and accuracy as well as stiffness depend strongly on the actual position of the platform.

poor. In the design shown in Fig. 18.36b the 1-DoF hinges of the powered upper arms and the parallelograms in the lower part prevent any rotation of the platform from the outset. Devices of this type are used mainly for pick-and-place tasks; the "Delta 4" is reported to place up to 150 randomly arranged pieces of chocolate per minute into the chocolate box.

The main strong point of the "Delta Robot" (b) is the extremely high possible acceleration of the (very light) platform which is said to reach 4 g. Here the angular DoF are stopped by using 1-DoF hinges in the upper arms and by having two struts arranged in parallelogram-fashion below [15].

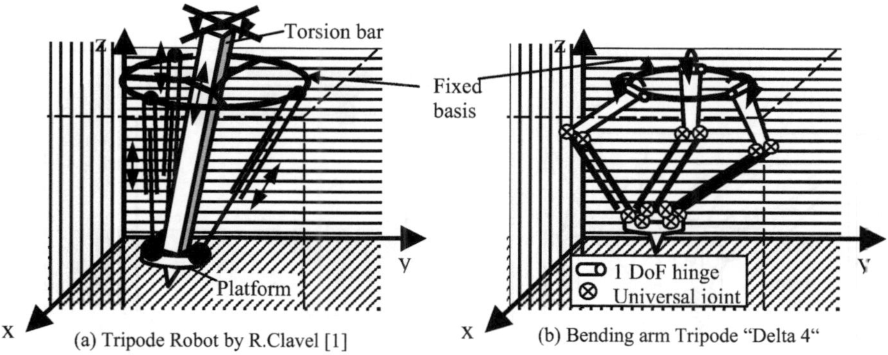

Fig. 18.36 Two examples of "Tripode" arrangements with 3 DoF. Both of them are very fast. Version (a) features a passive central strut which stops all angular DoF. Its useful workspace is exceptionally large but stiffness in the x,y directions is poor.

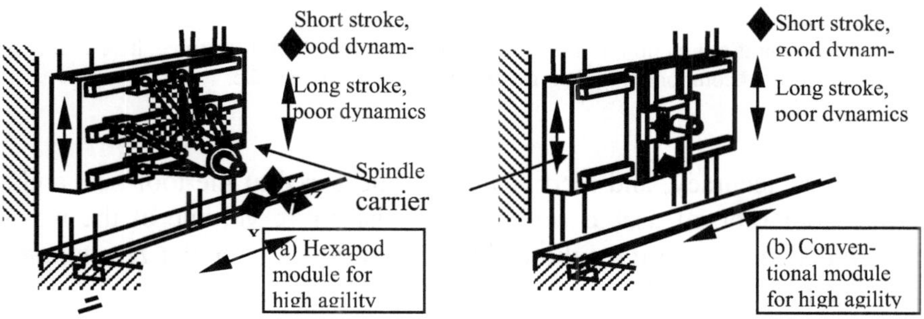

Fig. 18.37 To improve local agility of large machine tools, small, highly dynamic modules can be fixed to the spindle carrier in place of the normal, heavier work spindle. If three or more "fast" axes are required a Hexapod (a) or Tripod module will be appropriate. If two axes x,y are sufficient a conventional solution (b) is less expensive to realize.

18.3.1.3 Present Applications

The first "Hexapod," built by Giddings & Lewis, was shown to the amazed world in 1994 at the IMTS in Chicago. It had been praised in advance for exhibiting much higher stiffness (the struts carrying the platform are loaded only in the axial direction; no bending moments), higher dynamics (due to the principle of "Parallel Kinematics") and higher accuracy than conventional machine tools. While these predictions held t rue to some extent, it suffered severely from vibrations of the fixed structure as well as the moving struts; from the uneven distribution of most important parameters

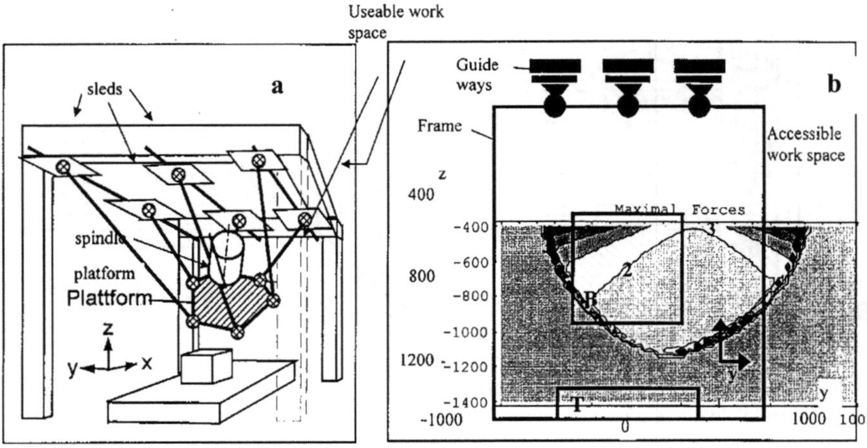

Fig. 18.38 a Cross-sectional view of the Hexaglide at the center position $x = 0$: Frame size, position of the powered guide ways, accessible work space, usable workspace, and ratio of feed drive force to force at the tool tip. b *Hexaglide*: A positive result of the uneven parameter distribution is a direction-and-position-dependent, very high feed force, here exemplified by the vertical component F_z near the bottom end of the *Z* stroke.

like stiffness, accuracy, and feed power (all measured at the tool tip), and from the poor ratio of usable workspace to the entire volume of the machine. Actual data regarding these features are shown for the *Hexaglide* [16] in Figs. 18.37 and 18.38 a.

In the mean time, additional problems have arisen. It turned out to be very difficult to calibrate these machines due to the lack of an "obvious" measuring basis; a repeatable positioning accuracy of ±10 μm is considered to be about the ultimate that can be attained at present. Another problem came up with the spherical joints. It looks almost impossible to design them according to the "ideal" specifications calling for high stiffness, large pivot angles of up to ±90°, and all rotary axes meeting in a single point within a very narrow margin. Thus, at present a rough evaluation looks like the one shown in Table 23.1.

Successful applications in industry have been reported about Hexapod machine tools (principally see Fig. 18.35a in die sinking where their 5-axis capability and their high dynamics have proven to be advantageous. Accuracy was not a problem as the workpieces rarely had to be more precise than, say, ± 0.025 mm. Another known field of successful application is very fast pick-and-place tasks performed mainly by tripod machines (see Fig. 18.37). They excel in very high dynamics due to the small inertia of the struts and joints that control the position of the grab-and-hold fixture at their tip.

Table 18.1 Strong and weak points of Parallel Kinematic Machines (PKMs).

Strong points of PKMs	Weak points of PKMs
Good dynamics due to small inertia	Uneven parameter distribution
Up to 6 equivalent degrees of freedom	Smaller workspace than in Cartesian machines
Potential for low cost due to many repetitive parts (e.g., struts and joints)	Demanding and time-consuming calibration
Lends itself well for modular design	Optimum set of design parameters is hard to find
	Poorer accuracy than in Cartesian machines

Contrary to early enthusiasm, industry has been quite skeptical vis-à-vis these machines recently. This dragging acceptance may have to do at least as much with the slow process of their reaching operational maturity as with the weak points shown in Table 23.1. Also, in trying to convince industry, the developers of PKMs (at least half of them are Universities) have tested their devices mostly in classical machine-tool applications (where often they could not perform to the best of their abilities), but shied away from novel applications. A few suggestions for such novel applications are shown in the next section.

A Look into the Future

The following suggestions for future, extended applications of PKMs are restricted to the Hexaglide scheme (see Figs. 18.36b and 18.37) because most of the author's experience has been gained with these.

The basic paradigm proposed here is to turn the so-called weak points of a PKM into strong points by tailoring the production problem at hand in such a way as to match best the abilities of the PKM.

Two examples may illustrate this point. In the first application we realize that the feed force available at the tool tip exhibits a pattern as sketched in Fig. 18.38a. Its vertical component is strong in the lower part, but weak in the upper part of the accessible workspace; for the horizontal component, the contrary is the case. In Fig. 18.39b, the "f" numbers indicate the ratio of the vertical spindle force F_z to the feed drive force F_x at various distances h to the vertical end point. Clearly over a distance of more than 80 mm, the force available is at least twice as strong as the drive force F_x. This suggests this region to be very well suited to perform "power operations" like vertical drilling and even riveting and bending operations.

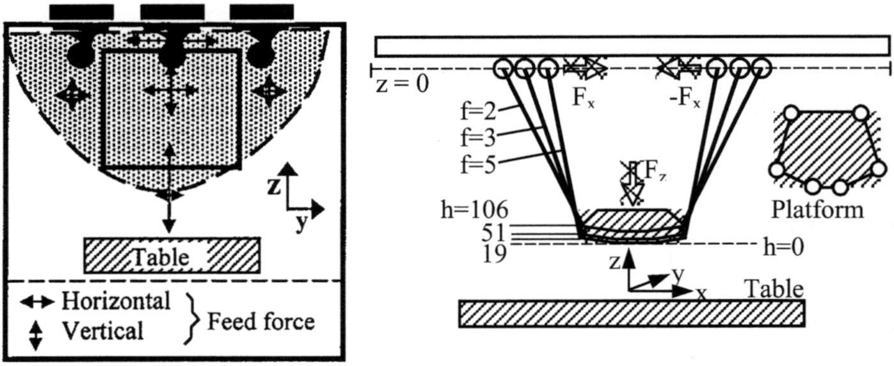

(a) Uneven feed force distribution in a "Hexaglide"

(b) Vertical feed force component grows rapidly as the bottom end of the z stroke is approached ($h \rightarrow 0$). $F_z = f \times F_x$.

Fig. 18.39 "Hexaglide": A positive result of the uneven parameter distribution is a direction and position dependent, very high feed force, here exemplified by the vertical component F_z near the bottom end of the z stroke.

A second example is to suggest how the high geometrical versatility of PKMs could be applied to extend their range of process capabilities. Some features may have to be added to this end as shown in Fig. 18.39. A secondary workspace centered about an axis orientation that is turned by 90° with respect to its usual vertical position (A); an automatic pallet/fixture change (B); and ways to attach a variety of work heads to the platform [(C): in C1 a drilling spindle, and in C2 a laser head are attached to the platform].

It will be shown in the next chapter that versatility is expected to become the longer, the more it is a key asset of successful manufacturing. It is fairly safe to assume that, once the hurdle of basic acceptance of the PKMs in industry has been overcome, their very high versatility as claimed in Table 18.1 (which also will improve further) will make this type of machine a favorite for a lot of production tasks.

18.3.1.5
It's the Match Between Task and Tool that Counts

In principle, the modern tools for designing and operating complex machines make feasible almost any configuration which is compatible with the laws of physics and geometry. But all of them still have their strong and weak sides; the ideal machine tool for any arbitrary production task belongs to utopia. What can be done, however – and it pays off! – is to devise and analyze novel concepts. It is important, however, to scrutinize not only the "tools" but also the tasks we want them to perform. Only if both sides –

the tool and the task – are shaped such as to match in an optimum way will they be fully successful.

18.3.2
Making a Product from A to Z in a Single Setup

18.3.2.1
Individual Parts

It is well known that, even in highly automated manufacturing, almost all products spend 80 to 90% of their "time of becoming a product" in an idle state. A major contributor to this idle time is the changeover between different manufacturing steps (which require different equipment), including multiple clamping, intermediate storage, and auxiliary functions like deburring, cleaning, quality inspection, and transportation in general. Avoiding such wasted time by finishing a product or parts of it in as few steps as possible is a very effective way to save costs and time.

As most products, however, still require multiple steps for their completion, manufacturing engineers have been looking for ways to minimize the number of changeovers. Options to this end are as follows:

Select a manufacturing process capable of machining a piece all the way from the raw blank to the finished part. Examples include: Injection molding of loosely toleranced parts; manufacturing drills out of tool steel bars by power grinding and finish grinding in a single clamping; cold forming of screws.

Combine several processes on a single machine tool. Example 1: *Turning, drilling and milling on a lathe* with powered tools; see Fig. 18.42. Performing all of these operations in a single setup often not only saves time

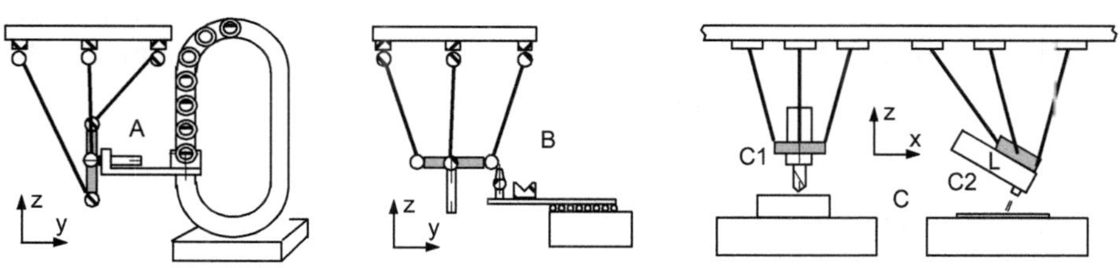

Fig. 18.40 Extended applications of a Hexaglide PKM in which use is made of its high geometric versatility. In Figure A, a secondary work space is assumed to exist in which the principal spindle orientation is turned by 90° with respect to its usual vertical position. This would allow, e.g., to change tools as shown; to some extent even "five side machining" of cubic workpieces would be possible, too. In Figure B an external slide table brings fixtures into the work space which are used in punching, drilling (C1), or laser cutting (C2). L is the laser head.

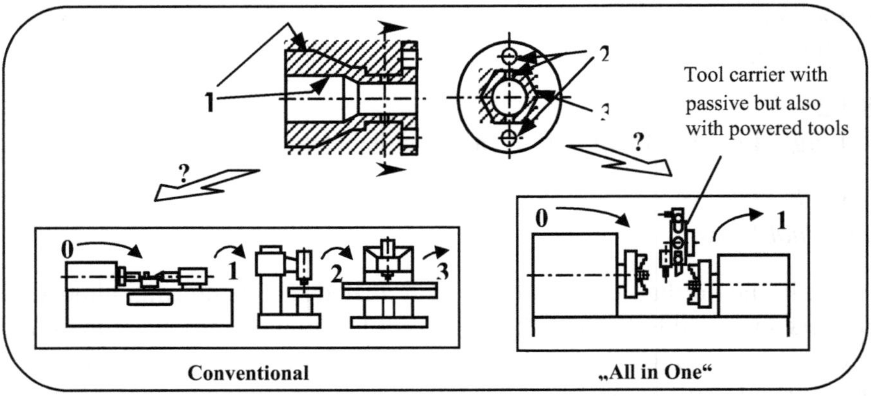

Fig. 18.41 The part shown in he upper part of the figure requires turning, drilling, and milling operations. In conventional manufacturing they are carried out on 3 different machines, thus calling for 3 separate clampings. A modern lathe with powered tools (here shown also with 2 spindles / chucks) performs all of these operations in a single set-up.

but also results in very high quality since all operations rely on one and the same axis (reference) system. Since all processes combined here are of the same type, i.e., cutting with geometrically defined cutting edges, their combination shows up very frequently in actual workpieces.

Example 2: *Sheet metal technology* has gained impressive ground with the advent of powerful software packages for 2-D and 3-D forming; with lasers coming up for fast and accurate cutting and trimming of arbitrary shapes and for fine welding, and with the development of materials optimized specifically for drawing and bending. "Sheet-metal-machining centers" have been developed capable of punching, nibbling, laser cutting, laser welding, and bending. The appropriate devices for positioning and also for assembly operations like riveting are there, too, ready for arbitrary uses. Thus, e.g., metal boxes complete with cover, hinges, and decorative cutouts are readily made on such machines in one setup.

18.3.2.2
Complete Products

The combination of several processes on a single machine tool suggested above for getting rid of idle time is a highly versatile scheme, but it suffers from the poor utilization of the equipment. Since in most cases only one operation can be performed at a time, all other equipment is idle. Therefore it is suited primarily for flexible, small-scale production rather than for mass production.

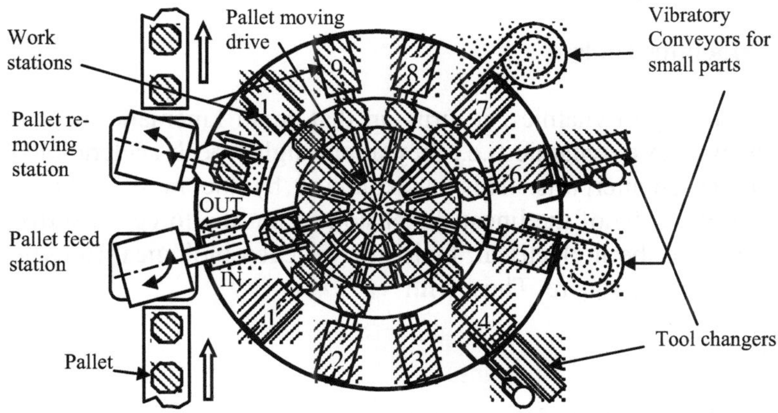

Fig. 18.42 Top view of a fully automatic rotary dial transfer machine. It works at a fixed rhythm; the pallets are all advanced simultaneously by one station at a time by the central drive unit.

The two desirable features: To complete the product in one setup, and having all equipment permanently at work, are realized in automatic *transfer lines* in which, preferably on a common base structure, all process stations are arranged in the appropriate sequence. As an example, a fully automatic rotary-dial transfer machine is shown in Fig. 18.41. Its strictly modular design allows an exchange of work stations, including all connections to the machine base (NC data, cutting fluids, etc.), and auxiliary equipment like tool changers and small-part supply in a very short time. This makes them quite flexible also for medium-size batch production, and even random production is possible to some degree. However, this high flexibility is possible only since the precision pallets used represent a reliable standard interface between the (random) workpieces and the (standard) work places. Thus, this type of machining center is about to become much more generally accepted than a short time ago.

More flexible approaches to combine diverse operations at random in a single machine are suggested from the side of the PKMs. In addition to the examples mentioned previously, another option is shown in Fig. 18.42. In a Hexaglide, guide-way modules have been added on both sides of the main work place. Additional platforms *nr.* 2 to 5 including struts and drives can be stored there which are fully operational and can be moved into the main work area in a few seconds. This solution is very flexible, but expensive, since all platforms except one (here: *nr.1*) are idle.

The 10 workstations can be equipped with any production unit fitting into the space and time frame allowed. They can be supplemented at random in a modular way by tool changers, part supply units, or whatever is needed.

18.3.3 Two-Stage Machine Tools

In the search for high dynamics, it turns up that small kinematic mechanisms are always more dynamic (i.e., they attain higher peak acclerations) than their larger counterparts. This is explained by the mass *m* being proportional to the cube of the linear dimensions *d* but the force *F* exerted by any power drive, be it electrical, hydraulic, or even biological (muscle force) growing with its square. Thus, with

$$m \sim \rho . d^3; \; F \sim \sigma . d^2$$

(ρ = mean density; σ = drive force proportionality factor), the scale law for the acceleration *a* is

$$a = F / m \sim \frac{\sigma * d^2}{\rho * d^3} \sim 1/d$$

I.e., the peak acceleration in a powered device is proportional to the inverse of its linear dimensions.

In conventional machine-tool concepts, the designer has to decide whether to strive for a large work space, meaning a large machine tool with rather poor dynamic performance, or a small, agile machine with a limited work space. Recently designs have appeared on the market where this controversy is solved by fixing a small, versatile unit (for high agility) on a large substructure for large strokes. While a known solution employs a

Fig. 18.43 The Hexaglide type PKMs lend themselves readily to become very versatile manufacturing machines by adding arbitrary production modules. In this figure, e.g., additional guide ways have been added on both sides of the main work unit. They can be used as storage areas for additional platforms which can be moved into working position in a few seconds. The figure suggests a combination of (1) a vertical drilling/milling spindle, (2) a clamping device, (3) a horizontal drilling/milling spindle, (4) a laser head, and (5) a bending tool.

Tripod as the fast unit as sketched in Fig. 18.43a, solutions on a more conventional basis (Fig. 18.44b) should yield similar results.

18.3.4 Conclusion

It is not by accident only that the developments shown here have appeared only recently, say, in the course of the last 10 to 12 years. Some basic ideas, however, like the principle of PKMs, are much older. This one has found widespread application in aircraft and truck simulators for more than 30 years. The big obstacle in those days to their application in machine tools, which are the most demanding products regarding the combination of speed and accuracy, was the lack of control power. Only the appearance of computers powerful enough to perform, e.g., the on-line coordinate transformations necessary for controlling a Hexapod, has made their actual realization possible.

One of the fascinating aspects of these solutions for the design engineer is the fact that all of them rely on very simple principles of logic and physics: The six-point theory, a branch of rigid-body kinematics; the management rule, "Finish off immediately what you have started"; and elementary physics. This may encourage the reader to trust his own intuition and take the fascinating risk of shaking the world by genuine imagination.

Literature

[1] Mitsuishi M, et al (1997) An open architecture CNC CAD-CAM machining system with database sharing and mutual information feedback. Annals of the CIRP 46 (1):269–274

[2] Merritt HE (1965) Theory of self-excited machine tool chatter. ASME J Eng for Industry pp 447–454

[3] Mitsuishi M, et al (1992) Real-time machining state detection using multi-axis force sensing. CIRP Ann 41 (1):505–508

[4] Mitsuishi M, et al (1996) A practical machining condition determination strategy using multi-axis force information. CIRP Ann 45 (1): 373–376

[5] Tobias SA (1961) Schwingungen an werkzeugmaschinen. Carl Hanser Verlag, München

[6] Hatamura Y, et al (1993) Development of an intelligent machining center incorporating active compensation for thermal distortion. CIRP Ann 42 (1): 549–552

[7] Mitsuishi M, et al (1999) Networked manufacturing with reality sensation for technology transfer. CIRP Ann 48 (1): 409–412

[8] H Shinno et al (1992) Structural configuration and performances of machining environment-controlled ultra-precision diamond-turning machine "capsule." CIRP Annals 41 (1):425–428

[9] K. Sawada, Y Takeuchi, and T Sata (1995) Development of ultra-precision machining center with closed-loop structure and its control. CIRP Annals 44 (1): 369–372

[10] Y Hatamura et al (1993) Development of an intelligent machining center incorporating active compensation for thermal distortion. CIRP Annals 42 (1):549–552
[11] Y Hatamura et al (1995) Actual conceptual design process for an intelligent machining center. CIRP Annals 44 (1):123–128
[12] M Nakao et al (1996) Development of an intelligent face grinding machine to fabricate ultra-flat surfaces on thin, brittle substrates. CIRP Annals 45 (1):397–400
[13] M Nakao et al (1997) A self-deformable face grinding machine to realize some 0.1μm flatness for 8-inch substrates. Proc. ASPE '97:238–241
[14] K Matsumoto et al (2000) Actively controlled compliance device for machining error reduction. CIRP Annals 49 (1):313–316
[15] Merlet JP (1997) Lex robots paralleles. Editions Hermès, 2nd edn, Paris
[16] Rehsteiner F, Hebsacker M (1998) Hexaglide – Erwartungen und erste Erfahrungen mit einem Sechsbeiner. Tagungsband Chemnitzer Parallelstruktur Seminar IWU, Chemnitz

Chapter 19
Reconfigurable Manufacturing Systems*

Y. Koren, U. Heisel, F. Jovane, T. Moriwaki, G. Pritschow,
G. Ulsoy, H. Van Brussel

19.1
The Challenge

Manufacturing companies in the 21st Century will face unpredictable, high-frequency market changes driven by global competition. To stay competitive, these companies must possess new types of manufacturing systems that are cost-effective and very responsive to all these market changes. Reconfigurability, an engineering technology that deals with cost-effective, quick reactions to market changes, is needed. Reconfigurable manufacturing systems (RMS), whose components are reconfigurable machines and reconfigurable controllers, as well as methodologies for their systematic design and rapid ramp-up, are the cornerstones of this new manufacturing paradigm.

The need and rationale for RMS arises from unpredictable market changes that are occurring with increasing pace during recent years. These changes include:

- increasingly frequent introduction of new products,
- changes in parts for existing products,
- large fluctuations in product demand and mix,
- changes in government regulations (safety and environment), and
- changes in process technology.

These changes are driven by aggressive economic competition on a global scale, more educated and demanding customers, and a rapid pace of change in process technology [1]. These drivers reflect a new balance among economy, technology, and society. To survive in this new manufacturing environment, companies must be able to react to changes *rapidly* and *cost-effectively*.

---

* This chapter has already been published as a CIRP Keynote Paper:
Koren Y, Heisel, U et al (1999) Reconfigurable manufacturing systems. Annals of the CIRP 48 (2):527–540

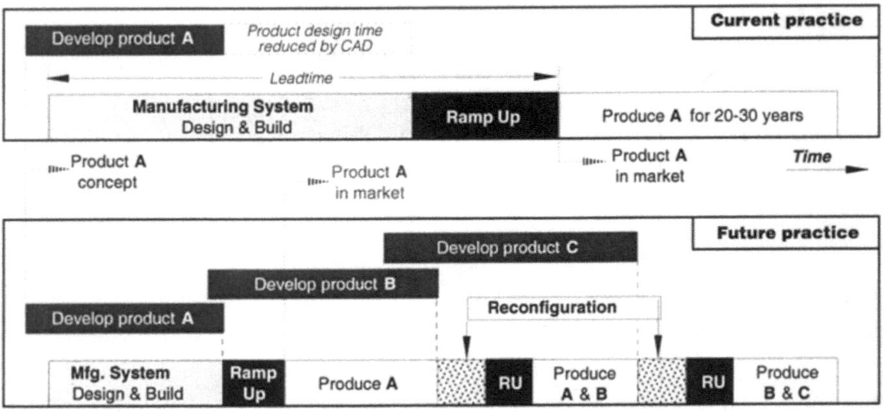

Fig. 19.1 While product-development time was reduced dramatically by CAD, nothing equivalent was done with the manufacturing system (top). An increase in the frequency of new-products introductions requires shortening the manufacturing-system-design time, and enabling its adaptation to production of new products through rapid reconfiguration.

To cope with the short windows of opportunity for introduction of new products, computer-aided design (CAD) has dramatically reduced product development times during the last decade (Fig. 19.1, top). However, such design methodologies do not exist for the manufacturing system itself, and therefore its design time remains lengthy. Manufacturing system lead-time (i.e., the time to design and build or reconfigure the manufacturing system, and to ramp-up to full-volume, high-quality production) has now become the bottleneck.

Brief windows of opportunity can be captured, along with major economic savings, if the lead-time of manufacturing systems can be reduced. Reduced lead-time can be achieved through the rapid design of systems that are created from modular components, or by the reconfiguration of an existing manufacturing system to produce new products, as depicted in Fig. 19.1, bottom [2]. In order to produce new products and accommodate required changes in existing products, new functions must be added to the manufacturing system through reconfiguration. This type of reconfiguration (i.e., adding manufacturing *functions*) is also needed for accommodating government regulations and integrating new process technology (such as new sensors, more reliable machine elements, etc.). Many reconfiguration periods will occur during the lifetime of the system. Short ramp-up (RU) becomes critical to successful reconfiguration.

A different type of reconfiguration is needed to cope with the large fluctuations in product demand and mix caused by the new market conditions. This type of reconfiguration requires rapid changes in the system production *capacity*, namely system scalability. In summary, a responsive system whose production capacity is adjustable to fluctuations in product demand, and whose functionality is adaptable to new products, is needed. Current manufacturing systems are not able to meet these requirements dictated by the new, competitive environment.

19.2 Types of Manufacturing Systems

Most manufacturing industries now use a portfolio of dedicated and flexible manufacturing systems to produce their products:

19.2.1 Dedicated Manufacturing Lines

Dedicated manufacturing lines (DML), or transfer lines, are based on inexpensive fixed automation and produce a company's core products or parts at high volume (see Fig. 19.2). Each dedicated line is typically designed to produce a single part (i.e., the line is rigid) at a high production rate achieved by the operation of several tools simultaneously in machining stations (called "gang drilling"). When the product demand is high, the cost per part is relatively low. DMLs are cost effective as long as demand exceeds supply and they can operate at their full capacity. But with increasing pressure from global competition and over-capacity built worldwide, there may be situations in which dedicated lines do not operate at full capacity.

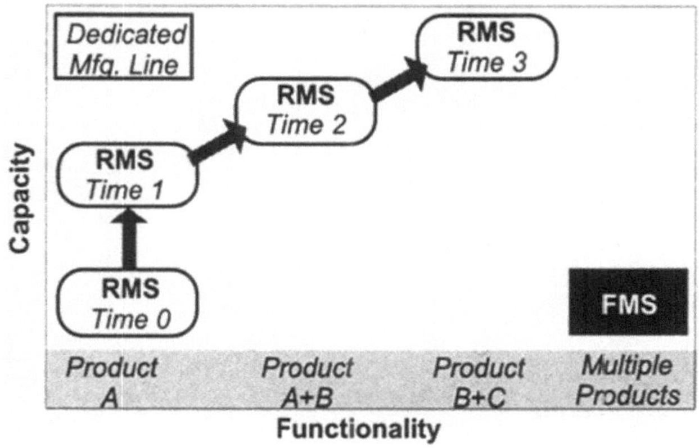

Fig. 19.2 Both DML and FMS are static systems, while an RMS is a dynamic, evolving system.

19.2.2 Flexible manufacturing systems

Flexible manufacturing systems (FMS) can produce a variety of products, with changeable volume and mix, on the same system. FMSs consist of expensive, general-purpose computer-numerically-controlled (CNC) machines and other programmable automation. Because of the single-tool operation of the CNC machines, the FMS throughput is lower than that of DML. The combination of high equipment cost and low throughput makes the cost per part relatively high. Therefore, the FMS production capacity is usually lower than that of dedicated lines and their initial cost is higher as depicted in Fig. 19.2.

The comparison between the two systems, shown in Table 19.1, identifies key limitations in both types of systems.

The challenge of coping with large fluctuations in product demand cannot be solved with dedicated lines that are not scalable. DMLs are not scalable because they are not designed for variable cycle times. Therefore, quite often, the available production capacity remains largely underutilized. A recent study carried out on a manufacturer of components for the car industry has shown that the average utilization of the transfer lines available was only 53% [3]. The reason for this low average utilization is that some products being in the early stages, or at the end of their life cycle, are required in low volumes. Even products in the maturity phase do not always reach the production volumes forecast at the moment of the design of the Dedicated Manufacturing Line.

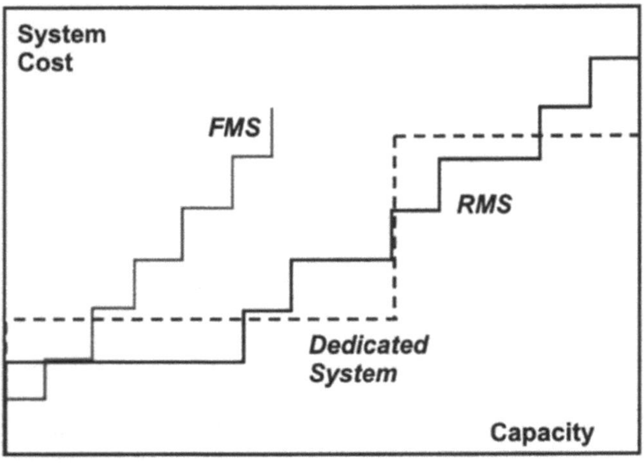

Fig. 19.3 Manufacturing-system cost versus capacity (or production rate).

Conversely, this challenge is theoretically met by flexible manufacturing systems that are scalable when designed with multi-axis CNC machines that operate in parallel. Despite this advantage, however, a recent survey showed that flexible systems have not been widely adopted, and many of the manufacturers that bought FMSs are not pleased with their performance [4].

19.2.3 The High Cost of FMS

The high cost of FMS is one of the major reasons for the low level of acceptance or satisfaction with FMS. Why is FMS expensive? Unlike DML stations, CNC machines are not designed around the part. Rather, general-purpose CNCs are built before the manufacturer selects machines and before process planning is undertaken to adapt the machines and the process to the part. Since the specific application is not known to the machine builder, the flexible systems and machines are constructed with all possible functionality built in. This creates capital waste. It is also a common assumption that FMS should be able to produce (1) any part (within the machine envelope), (2) at any mix of parts, and (3) in any sequence. This approach increases cost since it *requires a parallel system structure for FMS* that utilizes high-power, general-purpose multi-axis CNCs with a very large tool magazine and multiple sets of tools – a very expensive solution.

19.2.4 RMS – A New Class of Systems

A cost-effective response to market changes requires a new manufacturing approach that not only combines the high throughput of DML with the flexibility of FMS, but also is able to react to changes quickly and efficiently. This is achieved through:

- Design of a system and its machines for adjustable structure that enable system scalability in response to market demands and system/machine adaptability to new products. Structure may be adjusted at the system level (e.g., adding machines) and at the machine level (changing machine hardware and control software; e.g., adding spindles and axes, or changing tool magazines and integrating advanced controllers).
- Design of a manufacturing system around the part family, with the customized flexibility required for producing all parts of this part family. (This reduces the system cost.)

Table 19.1 Comparison between DML and FMS.

DML	FMS
Limitations:	Limitations:
Not flexible	Expensive
for a single part	*machine focus*
Fixed capacity	Low throughput
not scalable	*single-tool machines*
Advantages:	Advantages:
Low cost	Flexible
Multi-tool operation	Scalable

Table 19.2 RMS combines features of dedicated and flexible systems.

	Dedicated	RMS/RM	FMS/CNC
Machine Struc-	Fixed	Adjustable	Fixed
System focus	Part	Part fam-	Machine
Scalability	No	Yes	Yes
Flexibility	No	Custom-	General
Simult. Oper.	Yes	Yes	No

As summarized in Table 19.2, a system with these features constitutes a new class of systems – an RMS**.** The RMS is designed to cope with situations where both productivity and the ability of the system to react to change are of vital importance. Three coordinates – *capacity, functionality, and cost* – define the difference between RMS and the traditional DML and FMS approaches. While DML and FMS are fixed in capacity-functionality, RMS capacity and functionality change over time as the system reacts to changing market circumstances (Fig. 19.2).

Responsive systems are created by providing an adjustable structure, scalability, and flexibility (that although "customized," provides all the flexibility needed for the part family). Cost-effective systems are created by part-family focus and customized flexibility (rather than general flexibility) that enables the operation of simultaneous tools.

In the System-Cost vs. Capacity plane, the DML is a constant at its maximum planned capacity, and then, for greater capacity, an additional, expensive line must be built. The FMS is scalable at a constant rate (adding machines in parallel), as depicted in Fig. 19.3. The RMS is scalable, but at a non-constant rate that depends on the initial design of the RMS and market circumstances.

The main components of RMS are CNC machines and Reconfigurable Machine Tools (RMTs) – a new type of modular machine with a changeable structure that allows adjustment of its resources (e.g., adding a second spindle unit). In addition to RMTs, reconfigurable controls that can be rapidly changed and integrated in open-architecture environment are critical to the success of RMS.

The definition of an RMS is, therefore, as follows:

An RMS is designed at the outset for rapid change in structure, as well as in hardware and software components, in order to quickly adjust production capacity and functionality within a part family in response to sudden changes in market or in regulatory requirements.

If the system and its machines are not designed at the outset for reconfigurability, the reconfiguration process will prove lengthy and impractical. Systems designed for reconfigurability do not exist today, nor do their design and reconfiguration methodologies. However, many of the enabling technologies that are the cornerstones for reconfiguration do exist across the world.

19.3 Technologies Enabling Reconfiguration

The common denominator for existing dedicated and flexible systems is their use of fixed hardware and fixed software. For example, only part programs can be changed on CNC machines, but not the software architecture or the control algorithms. Therefore, these systems, including CNC and FMS, are static systems and are not reconfigurable. Manufacturing systems designed at the outset for reconfigurability do not exist today. During the last few years, however, two technologies that are necessary enablers for reconfiguration have emerged: in software, modular, open-architecture controls that aim at allowing reconfiguration of the controller [5]; and in

	Fixed Machine Hardware	Reconfigurable Hardware
No software	Manual machines, Dedicated mfg. lines (DML)	—
Fixed control software	CNC machines, robots, Flexible mfg. systems	Modular CNC machines
Reconfigurable software	Modular, open-architecture controller	RMS
System configuration rules & economic modeling		

Fig. 19.4 Classes of manufacturing systems. RMS design not only combines reconfigurable hardware with reconfigurable software, but also includes systems perspective and economic modeling.

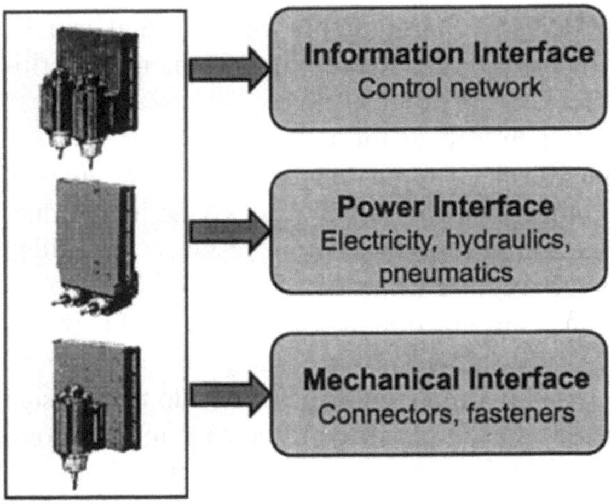

Fig. 19.5 Interfaces for machine modules.

machine hardware, modular machine tools that aim at offering the customer more machine options [6]. These emerging technologies show a trend toward the design of systems with *reconfigurable hardware* and *reconfigurable software,* as depicted in Fig. 19.4.

Reconfigurable hardware and software are necessary but not sufficient conditions for a true RMS. The core of the RMS paradigm is an approach to reconfiguration based on *system design combined with the simultaneous design of open-architecture reconfigurable controllers with reconfigurable modular machines* that can be designed by synthesis of motion modules. The ultimate goal of RMS is to utilize a systems approach in the design of the manufacturing process that allows simultaneous reconfiguration of (1) the entire system, (2) the machine hardware, and (3) the control software. The RMS paradigm will also create a new generation of reconfigurable machines that allow reconfigurations to achieve cost-effective scalability.

Unlike existing manufacturing systems that utilize fixed hardware and fixed software (e.g., CNC and FMS), the RMS will be designed through the use of reconfigurable hardware and software. With such design, the system *capacity and functionality* are not fixed but *change over time* in response to market demand, as shown in Fig. 19.2. This new type of RMS will allow flexibility not only in producing a variety of parts, but also in changing the system itself.

To fulfill the requirements of an open, modular machine structure, the modules and their interfaces must be specified in a well-defined manner. When examining a self-contained machine module, three main interfaces can be identified: mechanical, power, and information or control interface (Fig. 19.5). Only with the use of well-defined interfaces will RMSs become open-ended so they may be improved and upgraded rather than simply replaced.

Reconfigurable systems must be designed at the outset to be reconfigurable, and must be created by using hardware and software modules that can be integrated quickly and reliably; otherwise, the reconfiguration process will be both lengthy and impractical. Achieving this design goal requires an RMS that possesses the several key characteristics listed below:

- *Modularity.* In an RMS, all major components are modular (e.g., structural elements, axes, controls, software, and tooling).
- *Integrability.* Machine and control modules are designed with interfaces for component integration. The integrated-system performance is predicted based on a given performance of its components and the interfaces of both software and machine hardware modules.
- *Customization.* This characteristic has two aspects: customized flexibility and customized control. Customized flexibility means that machines are built around parts of the family that are being manufactured and provide only the flexibility needed for those specific parts, thereby reducing cost. Customized control is achieved by integrating control modules with the aid of open-architecture technology, providing the exact control functions needed.
- *Convertibility.* In a reconfigurable system the optimal operating mode is configured in batches that should be completed during one day, with short conversion times between batches. Conversion requires changing tools, part-programs, and fixtures, and also may require manual adjustment of passive degrees-of-freedom.
- *Diagnosability.* Detecting unacceptable part quality is critical in reducing ramp-up time in RMS. As production systems are made more reconfigurable and are modified more frequently, it becomes essential to rapidly tune the newly reconfigured system so that it produces quality parts.

Modularity, integrability, and diagnosability reduce the reconfiguration time and effort; customization and convertibility reduce cost. Therefore, these key RMS characteristics determine the ease and cost of reconfigurability of manufacturing systems. *A system that possesses these key characteristics has a high level of reconfigurability.*

We will discuss below the state of the art and later elaborate on methodologies aimed at achieving the objectives of:

- system-level design for RMS,
- reconfigurable machine design,

– reconfigurable control design in an open-architecture environment, and
– ramp-up time reduction for new and reconfigured systems.

19.4 State of the Art

Reconfigurable manufacturing is the latest development in the general field of computer-integrated manufacturing systems. One of the first firms to develop an integrated manufacturing system was Molins Company Ltd. In 1967 this company presented the "Molins System 24," a flexible and integrated system, developed by Mr. Williamson, showing a novel way to increase productivity. In this system the machining stations were linked by an automated handling system for workpieces fixed on pallets. Four years later, in 1971, Sundstrand developed the "Shuttle Car System," a rail-type pallet-transfer system on which workpieces flowed to and from the machining stations, located along the rail track. This system, however, was suitable only for long and variable machining times. At the Leipzig Spring Fair in 1972, Auerbach, a machine-tool factory, presented the manufacturing system "M250/02 CNC." Equipped with two three-axes machining centers, three two-arm changers, and one four-arm robot, this system enabled a five-face machining of prismatic parts. A central computer was used to control the machining centers, but the workpiece handling was done manually from a central operator's station. In 1977 the development of Flexible Manufacturing system Complexes (FMC), a test-factory consisting of modular machining units and assembly robots for the production of a whole range of parts within a given envelope, started in Tsukuba, Japan (see below). In the late 1970s the development of Flexible Manufacturing Systems started, providing the possibility to produce small batches of many different parts using a single manufacturing system. Group-structured production cells linked with automated material-handling systems emerged in the '80s (e.g., by Max Müller and Fritz Werner). In the early 1990s the idea of agile production systems had been pursued, enabling short changeover times between manufacturing of different products. Since the mid-nineties, the trend has been going towards Reconfigurable Manufacturing Systems, systems that are capable of being quickly adapted to changing market requirements by providing exactly the needed functionality and capacity at any time.

Advances in reconfigurable manufacturing will not occur without machine tools that have modular structures to provide the necessary characteristic for quick reconfiguration. However, the lack of machine-tool design methodology and the lack of interfaces are the major barriers that impede modularity [1,7,8,9,10,11]. Reconfiguration seems increasingly difficult the closer one gets to the ironware side because hardware interfaces are much more difficult to realize than software or control interfaces. While the latter is more a standardization issue, the hardware interface issue is difficult because of its inherent technical complexity.

As to the hardware modularity issue, there are very few documented research projects described in the literature that tackle the problem in a generic way. Perhaps the first sizeable attempt to solve the problem, and still a landmark achievement, was the FMC (Flexible Manufacturing system Complex provided with laser) project [12] launched in 1977 by MITI in Japan, and culminating in 1983 in a test factory built in Tsukuba, Japan. Although the project was carried out a long time ago, the fundamental concept is still valid. Figure 19.6 shows the conceptual design of this complex machining mechanism composed of modular machine units and assembly robots. An example of modularized machining cell is also shown in the figure. The machine modules were stored in a warehouse and assembled to fit the product to be produced. Upon completion of production, the machines were to be disassembled into their modules and stored. Modular assembly of the plant was completely task-driven. The FMC system was meant to produce a whole range of prismatic parts within a given envelope.

Another large-scale initiative was launched by the European Union (EU) in the early nineties. Based on a European Commission-sponsored report [13] on the state and future of the European machine-tool industry, it formulated a survival strategy for the European machine-tool sector. The report asserted that, if machine tools were designed and built modularly, then machine-tool builders could specialize in particular modules instead

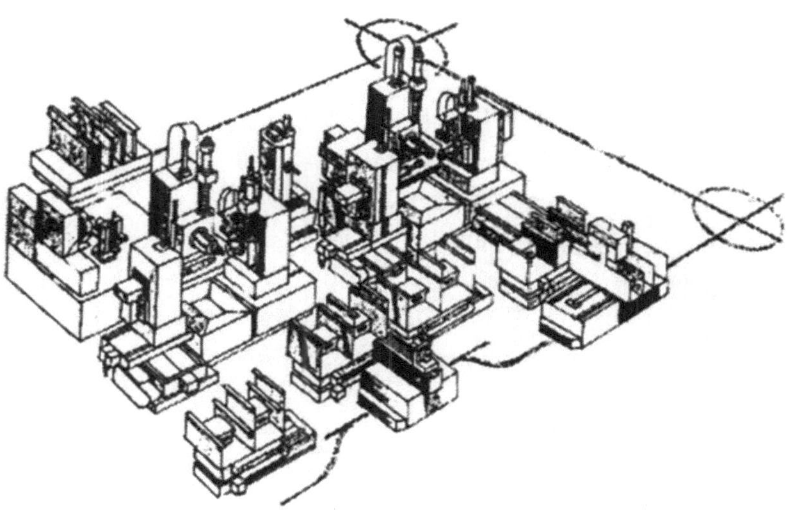

Fig. 19.6 The FMC project in Japan.

of in complete systems. System integrators would then build complete systems from the modules according to the specific needs of the users. This strategy requires splitting a machine tool into a set of autonomous functional units that can be plug-and-play interfaced to form complete systems for particular needs. Several European projects have been completed or are under development to achieve this goal. A few of them are discussed below.

The European MOSYN (Modular Synthesis of Advanced Machine Tools) project, led by Hanover University, looks at customer-specific configurations of modular machine tools. Another known project is KERNEL, which seeks to develop two different modular machine tools using equal-axis modules. The "Special Research Program (SRP) 467" at Stuttgart University, supported by the German research foundation, focuses on transformable business structures for multi-variant serial production. A sub-project within SRP 467 was entitled Reconfigurable Machining Systems, and assigned the goal of developing the basis for the realization of capacity and functional reconfigurability of machining systems. This project attempted to enable short-term adaptability of machine tools' capacity and functionality to the quickly changing production situation caused by turbulent environments [14]. The use of equal modules for different machines (see example in Fig. 19.7), and the design of interfaces, are important research issues in this project, as well as in another project entitled MOTION (Modular Technologies for Intelligent Motion Unit with Linear Motor and Axis Control) [15, 16].

In 1996, the Engineering Research Center of Reconfigurable Machining Systems (ERC/RMS) was founded at the University of Michigan by the National Science Foundation and 25 companies with the mission to de-

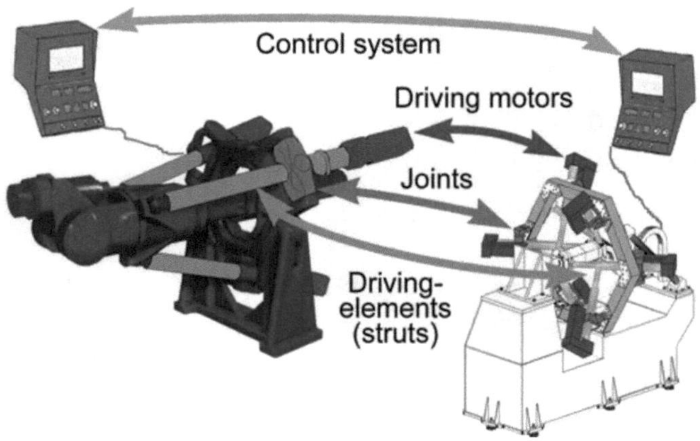

Fig. 19.7 Exchangeability of Modules.

velop the complete spectrum of RMS. The ERC/RMS has over 100 researchers that are developing RMS technology in three main areas: (1) Reduction of design lead-time of reconfigurable systems, (2) design of reconfigurable machines and their reconfigurable controllers, and (3) reduction of ramp-up time. The Center was awarded a patent for a reconfigurable machine tool [17]. The ERC/RMS takes a system perspective, not only in combining modular machines and controllers, but also in including the underlying methodologies for RMS design and operation. These include, for example, methods for system-configuration analysis and design, economic modeling, synthesis of reconfigurable machine tools, and calibration and ramp-up of RMS. An experimental RMS testbed serves as the verification tool for the developed technology. The aim of the center is to develop a scientific base for reconfiguration of machining systems. The science base will be applicable to other manufacturing domains.

Reconfigurable Modular Manipulator System, a related research initiative developed at Carnegie Mellon University [18], consists of plug-and-play compatible modules that can be assembled in a large number of different configurations to tailor the kinematic and dynamic properties of the manipulator to the task at hand. A similar concept is the cellular robot developed in Japan [19] that consists of modules from which a complete robot can be assembled. Finally, some projects carried out in the framework of the international Intelligent Manufacturing Systems (IMS) initiative also deal with problems of modularity and reconfigurability [20].

The recent Delphi study, Visionary Manufacturing Challenges for 2020, conducted by the USA's National Research Council, has identified reconfigurable manufacturing as first priority among “six grand challenges” for the future of manufacturing [21]. Various aspects of RMS are now under investigation by researchers across the USA (e.g., [22]).

Regarding issues of control, research efforts have been focused on open-architecture controls. On a global scale, the three most important initiatives in open-architecture control systems are the EU project, OSACA, and its German successor, HÜMNOS; the Japanese initiative OSEC; and the North American OMAC-TEAM project [5]. With the goal of specifying reference architecture for control systems, the OSACA project (collaboration among Stuttgart University, Aachen University, and industry) started in 1992. The main outcome from the OSACA project is the object-oriented design and specification of a vendor-neutral open architecture for machine-control systems. OMAC is an initiative driven by the desire to establish a set of application-programming interfaces (APIs) to be used by vendors to sell controller products and services to the aerospace and automotive industries.

In addition to exhibiting openness, controllers of future reconfigurable machines will be distributed and heterogeneous. The above-mentioned projects have not explicitly taken these issues into consideration, although they allow control of heterogeneous, distributed processes to some extent. Worth mentioning in this respect is the EU-sponsored project HEDRA (Heterogeneous and Distributed Real-time Architecture) [23], which uses the real-time kernel VIRTUOSO. Also the MOTION project, mentioned above, tackles synchronization and interpolation issues when combining several intelligent single-degree-of-freedom linear motion-control modules.

Several machine tool builders develop their current products as modularized systems to facilitate design and to offer customers a customized product at an affordable cost. Such modularized systems include horizontal and vertical turning centers, as well as transfer lines. However, designing a system with a combination of modules from different manufacturers requires standardization of the mechanical interfaces. Compared with the standardization of interfaces for information technology, such mechanical interfaces have not taken on an important effort to date [24]. A standardization of products across the range of manufacturers is advantageous for the use of the same modules for different machine tools, whereby machine quality can be improved, cost reduced, and reconfigurability of the machines enhanced.

To help assess the near-future (5-10 years) developments in manufacturing systems, a survey has been conducted by the UM ERC/RMS. In this study, experts in the field of manufacturing were asked to make predictions based on their knowledge of the manufacturing field, and to present the rationale behind their forecasts. This survey project has accomplished two main goals: (1) To examine the results to date associated with the use of existing manufacturing systems, such as flexible machining systems; and (2) to examine the potential roles, justifications, and enabling technologies for reconfigurable machining systems in future manufacturing facilities. As part of this second goal, the panel identified key enabling technologies needed to realize these benefits. The results of this study have been documented in an ERC/RMS report [4]. The main conclusion of the report is that, to achieve the required responsiveness, manufacturing should be viewed, designed, and optimized as a system (as a whole). Existing major barriers to this achievement are the lack of available tools and methodologies to analyze the trade-off among processes, equipment, life-cycle costs, and initial investment.

19.5 System-Level Design Issues in RMS

Unpredictable, frequent market changes already characterize the environment of many manufacturing companies. Changing orders cause altering requirements concerning the output capacity and the processing functions of the manufacturing systems. Reconfigurable manufacturing systems meet these requirements by offering a rapid adaptability of both their capacity and functionality to the new situation. This adaptability feature of RMSs also offers a short-term resetting of the manufacturing systems to produce different variants of the current products. Furthermore it provides system adaptability to new products and thereby guarantees a high long-term benefit-to-cost ratio [2, 25].

Similar to flexible systems [26, 27, 28, 29] RMSs are equipped with automated workpiece and tool supply. The structure of the supply systems influences significantly the productivity, part quality, and reliability of the RMSs [30, 31].

19.5.1 Life-Cycle Economics

If we take into account the entire life-cycle cost of a production system in an uncertain market, reconfigurable systems can be less expensive than flexible manufacturing systems or even dedicated manufacturing lines. The main factor that makes the RMS less expensive is that *RMS is installed with precisely the production capacity and functionality needed* and may be upgraded, in capacity and functionality, in the future, *exactly when needed.* Expanded functionality enables the production of new parts of the same part family on the same system, which in turn expands the lifetime of the system. RMS allows one to add the extra capacity exactly when required, which solves the under-utilized capacity problem of dedicated lines. RMS also allows the additional functionality to be added exactly when needed which saves the investment for unneeded functionality associated with FMS. Thereby the two types of waste that occur in dedicated and flexible systems are eliminated with RMS technology.

19.5.2 Definition

A system configuration is defined as a set of machines (including controls) and the connections among them. Machines may be either given or composed of modules. The definition of the *system-level configuration task* is as follows:

Given: (a) a part or a part family, (b) volume and mix, (c) libraries containing (i) a fixed set of machine modules (each described by shape, interfaces, kinematics, and stiffness), (ii) constraints regarding the assembly of these modules, and (iii) a fixed set of machines (each described by a set of kinematics, processes, etc.), and (d) constraints that describe the permitted sequences of processes on the part; build configurations that specify all these requirements, or detect incompleteness in the given sets of machines or machine modules.

The number of possible configurations is very large. An increased number of machines cause a combinatorial explosion of system configurations. For example, for six machines there are 54 possible configurations: a serial, a parallel and 52 hybrids (parallel-serial combinations). Each configuration has to be evaluated for throughput (function of reliability, machine speed, mix, and desired volumes), part quality, and cost. This situation makes the consideration of alternative configurations extremely difficult. Simulation software and throughput analysis tools are widely used by industry to design manufacturing lines. However, these tools do not account for part-quality issues. Hence, up-front trade-offs between quality and throughput could not be done concurrently. Novel evaluation algorithms that enable a quick, integrated, throughput-quality analysis of configurations were introduced in [30]. These algorithms, however, solve each demand scenario as an independent case, and do not optimize the system during its entire lifetime for different configuration scenarios. Work is underway to calculate the optimal scalable system configuration.

Furthermore, conducting only economical evaluation is insufficient in many cases and the dynamics of the system must be also considered. Ranking alternative configurations may also be done by other criteria such as the ability to react to change, environmental impact, or compliance with production systems already installed in the company. The Analytical Hierarchical Process (AHP) methodology and the ELECTRE methodology have various means of weighting the various objectives [31]. In [32], multi-criteria genetic algorithms are used to optimize machine-tool configurations at the conceptual stage. The optimization criteria are static stiffness, dynamic stiffness, and workspace.

19.5.3 Modular Structure

Reconfigurable manufacturing systems need a modular structure to meet the requirements for changeability [33], which is provided by a modular system structure [34]. The primary goal in developing RMSs is to develop machine modules, which can be quickly exchanged between different manufacturing systems. This exchangeability can be accomplished by equal structure of the machines and the control systems and the standardization of the interfaces combining the modules, which enables a short-term adaptability to market changes by reconfiguration of the manufacturing system.

To guarantee easy reconfigurability, not only the physical system must be updated, but also the management and control software must take into account the new characteristics of the plant. This is needed to ensure the proper flow of materials, tools, and information. It is worth mentioning that one of the main reasons for the failure of the early FMSs was the rather problematic supervision of the system. In RMS the supervisory program must be adaptable to different system configurations. A project called OpenFront that is conducted at the ERC/RMS is developing control software to manage new plant reconfigurations [35].

One of the simple countermeasures to cope with changes in the production volume is to add or remove machine tools in the machining system. Figure 19.8 shows an example of construction of reconfigurable transfer lines proposed and implemented by Toyoda Machine Works. The line includes five basic kinds of modularized base machines with different capa-

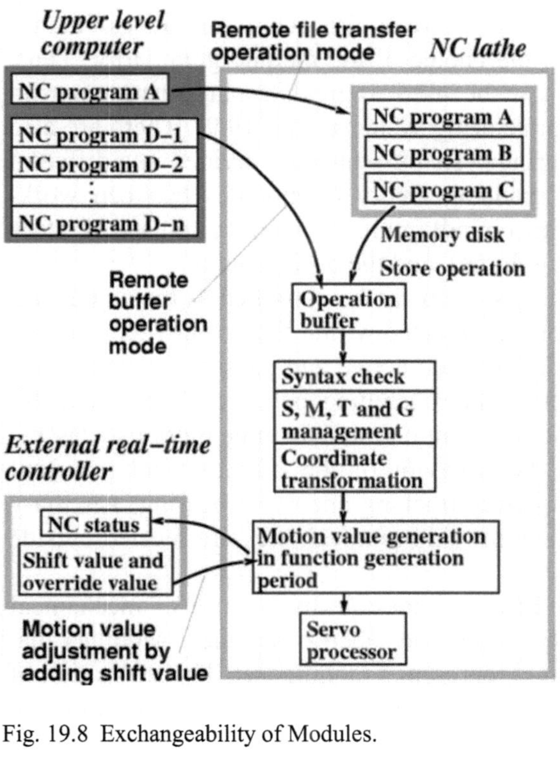

Fig. 19.8 Exchangeability of Modules.

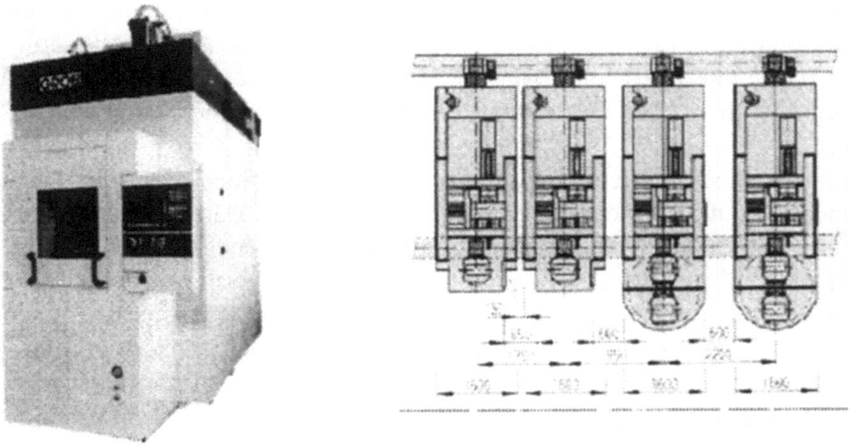

Fig. 19.9 Machining centers as modules of a large machining system.

bilities in cutting performance. Each machine is equipped with an automatic tool changer. Toyoda claims that flexible and modularized work clamping jigs and work transport units are the keys to success of this type of reconfigurable machining line.

The influence of the modular structure on the reconfigurability of manufacturing systems depends on the choice of the module granularity. Thus, for example, machine tools can be used as modules in constructing an RMS similar to flexible manufacturing systems (Fig. 19.9). There are also other examples for manufacturing systems, which are divided into different system levels to enable the exchangeability of modules between workpiece-dependent subassemblies (spindle units, workpiece transport), standard subassemblies (machine column), and the basic machine.

The functions of a machining system can be classified as main and auxiliary functions. Main functions are located within the power flux of a machining process and can be divided into active and passive functions. Active main functions characterize the machining power and represent dynamic functions, e.g., positioning motions and cutting motions. Passive main functions characterize the behavior of the machine and represent functions such as tool or workpiece tracking and force bearing. Auxiliary functions, such as supply functions, chip removal, etc., are located outside the power flux.

The rapid adaptability of RMSs is possible by the use of mechanical modules, control modules, hydraulic and electric modules. Modules that can be exchanged and integrated represent a modular set. The realization of modular sets requires a standardization of the interfaces connecting the modules.

19.5.4 Interfaces

To realize RMSs and their modularity, standardization of the module-connecting interfaces is required. The choice of the module granularity specifies the type of interface. Interfaces can be divided into international, national, and company-specific standardized interfaces. National and international tool and tool-holding fixtures like hollow shanks are examples for existing standardized interfaces. Guiding profiles, indexing modules, and clamping elements of pallet systems are also standardized, while drive elements and carrying devices are not, which handicaps the exchangeability of pallets between different systems.

The interfaces of a machining system can be subdivided into three types:

- The system interface determines the connections between machining systems, needed to combine several machines to a manufacturing system. Appropriate interfaces prescribe the logistic connection and the connection to further machining systems. This type of interface is standardized in order to combine different systems.
- The module interfaces represent interfaces between singular modules. They are standardized in order to integrate modules of different manufacturers into a machining system.
- Submodule interfaces determine the connections inside the modules and make it possible to assemble modules from sub-modules. This enables one e.g., to combine different driving motors (sub-modules) with a spindle system (sub-modules) to generate different spindle-unit-modules (modules). The national or even international standardization of these interfaces is not useful as it might handicap the manufacturer's progress and innovation in developing modules.

To configure machining systems, mainly the interfaces between the modules have to be considered. The configuration of single modules requires the consideration of the sub-module interfaces, and the configuration of manufacturing systems necessitates the contemplation of system interfaces.

Interfaces can be divided into mechanical interfaces and interfaces for data, energy, and auxiliary material transmission. Mechanical interfaces transmit forces and moments, align elements, and fix them together. The other interfaces care for the component supply with required media. Mechanical interfaces can be specified to dynamic and fixed interfaces. Dynamic connections can be opened and closed with quick grippers used for tool fixtures like hollow shanks, while fixed interfaces are used for connections only to be unlocked in case of system configuration.

19.6 Reconfigurable Machine Tools

The two core engineering methods needed for machine-level design are (a) a method for systematic design of modular, reconfigurable machines, and (b) design principles for open-architecture controller. The machine-design method should utilize a library of machine modules; each can provide a fundamental motion. This mathematical approach may generate new types of machines – Reconfigurable Machine Tools (RMT).

In the case of machining systems, the reconfigurable machines should perform such processes as turning, milling, drilling, and tapping, or a combination of these processes. In the broader manufacturing case, combined processes, such as machining, heat treatment, assembly, and metrology are to be considered. The discussions here, however, are basically limited to machining processes. The challenge in this case is to design an optimum RMS in terms of cost, productivity, part quality, and reconfiguration time.

The modular design of machine tools is a key enabling technology to reconfigurability, as the machining system can easily be reconfigured by simply removing, adding, or changing the constituent units or modules of the system or the machine. The concept of modular design of machine tools is already well known. However, recent development of new machine hardware is opening new possibilities to modular design of machine tools.

Figure 19.10 shows a schematic illustration of typical 3-axis modular-machining centers. The discussion here is basically focused on how to increase reconfigurability of this type of machine tool.

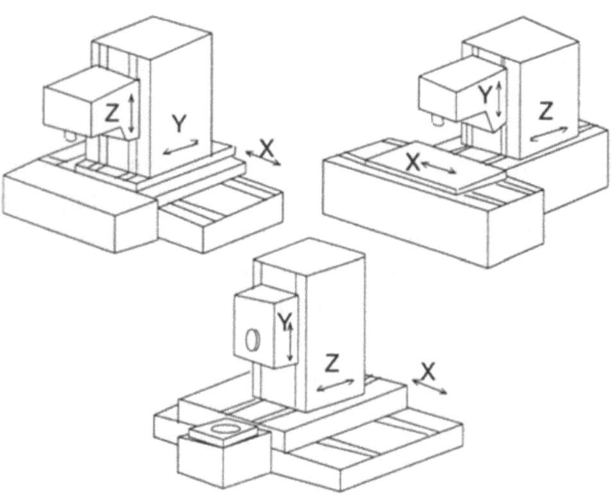

Fig. 19.10 Modular machining centers.

19.5.5 Variation of Products

The primary aim of an RMT is to cope with various changes in the products or parts to be manufactured. The following possible changes must be taken into consideration:

(a) Workpiece size
(b) Part geometry and complexity
(c) Production volume and production rate
(d) Required processes
(e) Accuracy requirements in terms of geometrical accuracy, surface quality, etc.
(f) Material property, such as kind of material, hardness, etc.

Reconfigurability for workpiece size. In order to cope with simple changes in the size of the workpiece, it will be enough to prepare machine units (modules), such as a column, table, spindle unit, etc. with different sizes. Reconfigurability is achieved by changing modules.

Reconfigurability for part geometry. In order to increase the machine functionality for geometric complexity of the parts, the number of axes-of-motion is increased by adding new motion units or by replacing one of the units with one having several degrees-of-freedom. Figure 19.11 shows a schematic illustration of a possible reconfiguration of a 5-axis machining

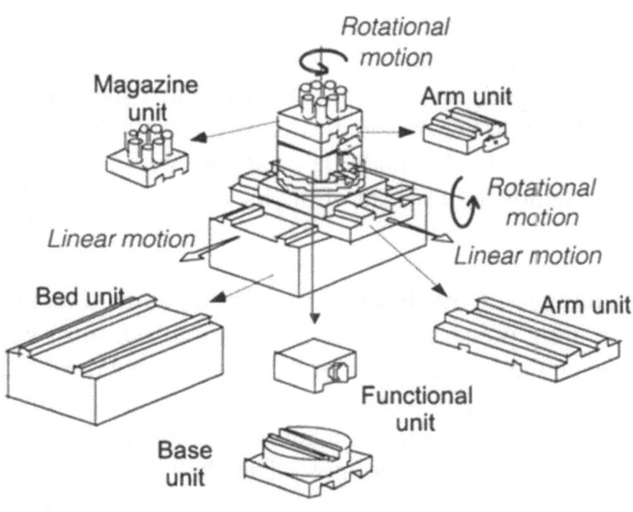

Fig. 19.11 Assembly of modules for a rotary axis

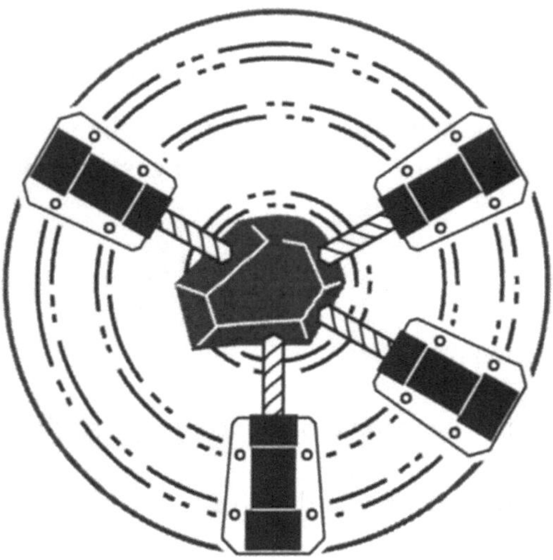

Fig. 19.12 Top view of a vertical turning-milling center with multiple spindles.

center. Some of the commercially available 5-axis machines are reconfigured by replacing a rotary work-table with a three-axis table, and the others by replacing the spindle head with a versatile spindle-head unit.

Reconfigurability for production volume and rate. In order to increase production rate, the capacity of the machine spindle unit can be changed from single-spindle unit to dual- or even multi-spindle unit. The multi-spindle unit is a very powerful tool to increase the productivity. Modularized spindle units with different speed ranges and horsepower are good examples of utilizing RMTs. An example of an RMT is depicted in Fig. 19.12, which shows a part mounted on a rotary table machined simultaneously with four spindles. The number of spindles may vary to accommodate the desired production rate. Each spindle is a module with a Z-axis. Two additional manual axes allow adjustment of the location of the spindle and its cutting angle.

Reconfigurability for changes in machining process. In order to cope with changes in the machining process, not only the cutting tool must be changed, but sometimes even the configuration of the machine tool must be changed as well. In some applications, not only turning, but also milling and drilling operations can be performed on a turning center utilizing a milling spindle, which replaces the fixed-tool post. The challenge is, however, adding, for example, a grinding unit to the turning center when needed. The machine in Fig. 19.12 (a top view shown) may be converted to a vertical turning center, in which the part is machined with multiple tools that may perform drilling, milling, and turning operations.

Reconfigurability for machining accuracy. RMTs consist of modules, each having its own interface. In such cases, the mechanical interfaces that are specified by geometric features, have associated tolerances. The undesirable addition and/or superposition of tolerance fields can have a negative effect on the machine accuracy after reconfiguration. It is not easy to increase the machining accuracy by simply changing the machine modules with other modules of higher accuracy, as the machining accuracy is determined by the combined motions of the tool and the workpiece, and also by accuracy of the relative arrangement of the modules and their interfaces. The machining accuracy is also influenced by the static and dynamic rigidities of the machine, and the thermal deformation of the machine as well [36].

Parallel-Kinematics in Reconfigurable Manufacturing Systems. Parallel-kinematic machines may form a special class of RMT. They can be categorized into different classes depending on the type of their strut – i.e., those with constant length and those with variable length – and on their drive design such as linear drives, ball screw drives, and rack-pinion drives

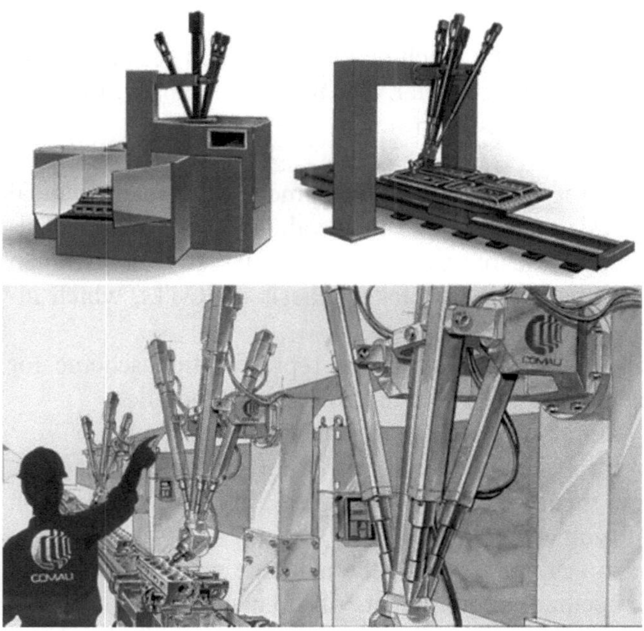

Fig. 19.13 Tripod-kinematic machines integrated into a machining line.

[37]. An exhaustive analysis of the fundamental technical principles of existing parallel-kinematic machines and/or prototypes has been presented in [38,39,40].

Most parallel-kinematics machines are realized as tripod or hexapod systems, which differ, by the number, the specification, and configuration of its struts. The characteristics of these kinematics result from their low-positionedmasses and the high-strut stiffness [24]. As these machines consist of simple and identical modules [41], it is possible to configure different machines from these modules.

The use of many equal modules implies the standardization of drive elements, joints, and module interfaces. As a result struts, joints, and drives become exchangeable between systems of different configurations, which enables the design of singular reconfigurable manufacturing modules. These parallel-kinematic machines can be used as particular modules in transfer lines for the machining of cylinder blocks with automated workpiece flow and as stand-alone C-styled or portal-styled laser centers for the machining of sheet metal (Fig. 19.13). It is also possible to realize flexible manufacturing systems for additive machining with this type of kinematics.

19.5.6 Challenges

The concept of modular design of a machine tool is not new, but it has not been widely adopted in practical use so far. The main problems with modular and reconfigurable design are as follows:

1. Design Methodology for RMT: The development of a mathematical framework for synthesis of RMTs and their validation is a major challenge. The Engineering Research Center at the University of Michigan is developing a mathematical theory for synthesis of RMTs, which includes [42]:
   - Development of a formal and a unified representation scheme for mechanical (kinematic, and structural) functions of modules.
   - Compilation of a library of machine modules (see Fig. 19.14).
   - A methodology for systematic synthesis of reconfigurable machine tools using screw theory for kinematics (see Fig. 19.14) and graph theory structural synthesis.
2. Interfaces: It is not so easy to assemble the machine module accurately enough to meet the accuracy requirements of the machine tool. The interfaces between the modules to be assembled must be standardized and accurately manufactured. Methodologies must also be developed to rap-

idly measure and adjust the alignment of the modules. Reduction of the static and dynamic rigidities at the interface is also a problem to be solved.

3. Module Autonomy: Most of the movable and drive units are supplied with electricity and are connected with the controller by wires. Some of them also require hydraulics and compressed air (see Fig. 19.5 above). The wiring and piping with the external energy sources are a nuisance. The location of hydraulic pumps, compressors, and accumulators can be an obstacle for reconfiguration of the machine units. It is desirable that each module should be autonomous and independently workable.

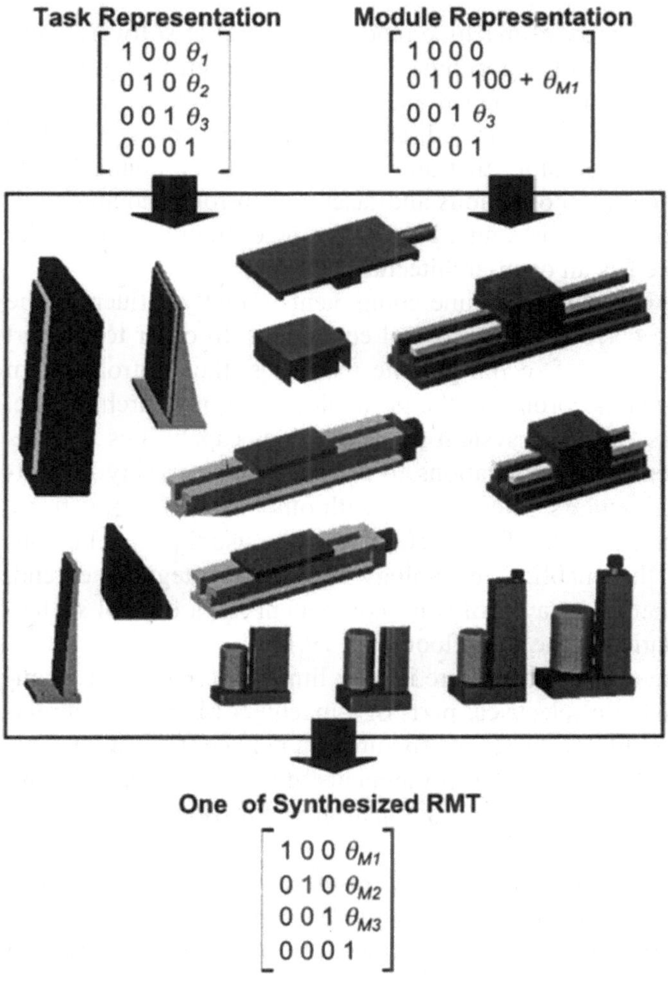

Fig. 19.14 The process of a formal design method for RMTs.

In conclusion, to cope with the changes in product variation and volume, the machining systems and the machine tools have been replaced or new machine tools have been introduced in the past. It is no more cost effective to do so, as the changing cycle is becoming shorter and the variations larger. The modularized machines and machine modules are expected to solve this problem by reconfiguring the machining system and machine tools. Effective utilization or re-use of modularized machine tools and machine units also contribute to sustainable society.

19.7 Control for Reconfigurable Machines in Open Architecture

Similar to building a library of machine modules, controller software components (e.g., servo-control algorithms, temperature-control algorithms, interpolators, etc.) are also catalogued and stored for re-use. The modules needed for the application are selected, and then configured by a method termed a "Control Configurator" that aids in integrating the controller for the selected machine (both continuous and discrete control), and automatically checks its real-time constraints. Finally the reconfigurable controller has to be implemented as an open-architecture control.

The modular design of the machine components directly influences the requirements for the corresponding control equipment. In order to support the modular construction of reconfigurable machines, the control system itself must be designed according to the principles of an open architecture. IEEE defines this as "an open system (that) provides capabilities that enable properly implemented applications to run on a wide variety of platforms from multiple vendors, inter-operate with other system applications, and present a consistent style of interaction with the user" [43]. The controller openness is the enabling technology needed to integrate, extend, replace, and reuse hardware and software components in a control system even after its installation at the shop floor (Fig. 19.15).

Due to the nature of hardware, there is only limited flexibility in reconfiguring the mechanical or electrical parts of a machine. Manual manipulation is required to modify an existing structure to a new configuration. This is different for the software-oriented components of the control. The immaterial nature of software allows the modification of functionality in a very dynamic way.

To obtain maximum integration flexibility, re-use and distribution of software modules must be achieved. A neutral *system platform*, which is independent from specific applications and which can be easily adapted to a specific hardware, is indispensable to fulfill these requirements [44]. The system platform encapsulates all specific parts of a control system in order

Capability	Meaning	Hardware Options	Software Options
Modular	decentralized structures are supported	standardized and distributed communication systems	distributed appl. modules
Inter-operable	components can co-operate		standardized data exchange
Portable	integration of components in different environments		standardized API
Scalable	Increased performance; topology can be modified	processor upgrade	parallel software tasks
Extendable	functionality can be enhanced	additional plug-in cards	additional appl. modules

Fig. 19.15 Capabilities of a control system for reconfigurable machines.

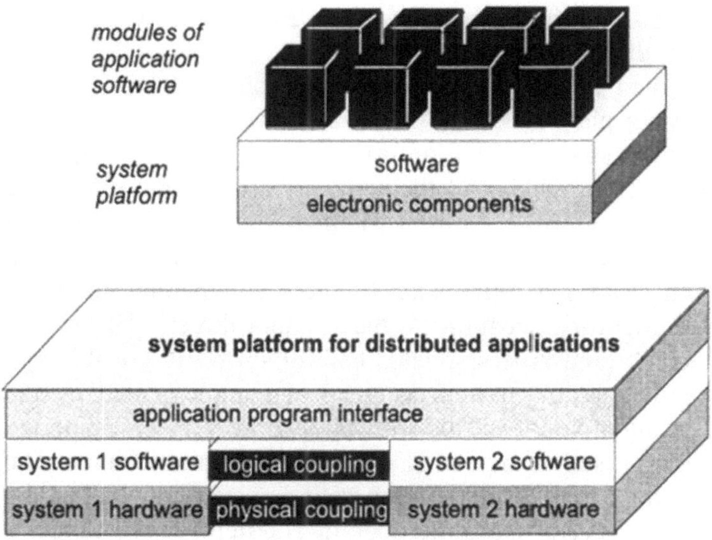

Fig. 19.16 Structure of the system platform (top) of an open-controller architecture (bottom).

to provide a neutral interface to the application software. The system platform is based on electronic equipment such as computer boards, plug-in cards, or intelligent I/O nodes connected via control networks. On top of the platform is located the system software which provides the functionality needed to support openness for the application software.

From the required capabilities for application modules (AMs), the basic functionality of the system software can be derived (Fig. 19.16). The desired portability of AMs leads to the need for the definition of a uniform application program interface (API) to access the system platform. This enables modules to be placed on any system platform that offers the API. This makes it also possible to change a certain configuration by repositioning a module in the system and by thus modifying a given topology. It implicitly demands that the system software has to shield the application software completely from the system hardware.

The system software for an open-control system defined as above has to contain at least three components:

- an operating system, or run-time system, to execute modules,
- a configuration system to combine modules into a running system at boot-up of the control, and
- a communication system to enable information interchange between modules on the basis of a standardized protocol.

Issues related to real-time limitations caused by the communications network bandwidth are discussed in [45].

Although combined control components on the physical level depend on the control system platforms (Fig. 19.16 bottom), the platform looks transparent to the user, since a set of uniform APIs is provided for the *application software*. The API hides all the hardware aspects and offers one homogeneous system platform even in distributed-control systems. Examples of such system platforms are CORBA of the Object Management Group (OMG) and DCOM of Microsoft. As these commercial platforms have no dedicated real-time capabilities, they cannot directly be used as platforms for machine- and process-related control software. For this reason, a specific system platform was designed and implemented by the European OSACA initiative [46]. This software can be very easily ported to different operating systems and can also support real-time features. The software was already ported to various operating systems including the MS-Windows family, several real-time operating systems (such as OS-9 and VxWorks), and different UNIX derivatives. For the physical coupling of distributed platforms, a wide range of communication standards (such as Sercos, Profibus, Ethernet) are available.

Using the above-described principles of a control-system platform with a system-independent API, application software can be designed and implemented in a very modular and flexible way. In order to support a concurrent design of the machine and the related control software, it is helpful that the software modules correspond to the components of the machine. In

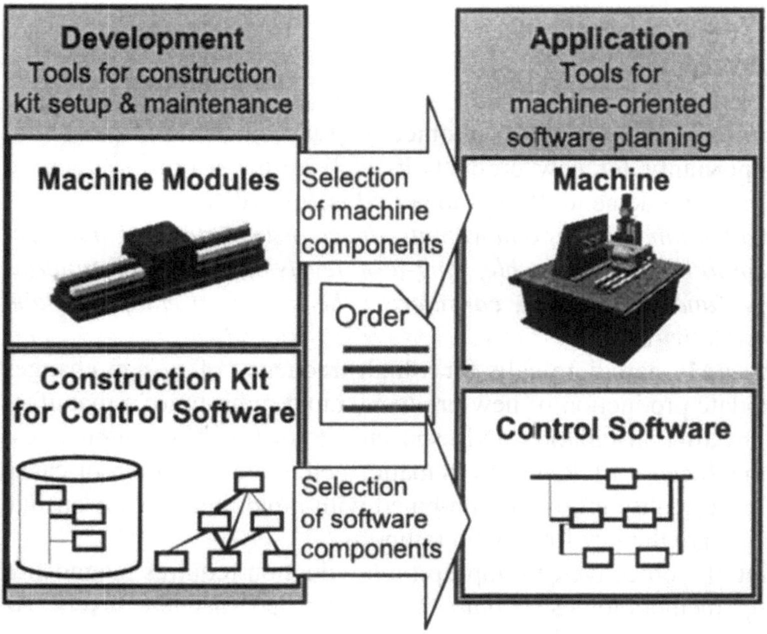

Fig. 19.17 Concurrent engineering of structure of machine and control software.

terms of object-oriented principles, it means that a machine object corresponds to a controller object offering the control functionality of the machine object (see Fig. 19.17).

Control software elements are deposited in a module library. They are encapsulated into independent units that contain both structural and functional information. This causes a high re-usability of the modules.

Based on this library, new configurations are generated by selecting the modules in a project and by associating the provided communication interfaces with each other. Automatic testing mechanisms prevent the generation of faulty configurations.

After the termination of the configuration process, the target code for the control is generated. This can be carried out according to vendor-neutral or vendor-specific formats, and can cover PLC and NC code.

Using a graphical configuration tool reduces the time for creating a control configuration by using libraries and building plans. Software can thus be produced at a lower cost. At the same time, the quality of the software increases because of its integrated testing mechanisms and its being based on models. First prototypes of such configuration tools have been designed and developed within the German projects MOWIMA and HÜMNOS [47].

19.8 System Ramp-Up

Ramp-up time reduction is a critical objective for responding to short windows of opportunity for new products as well as for scaling systems to cope with changing demand. Ramp-up period is defined as:

The period of time it takes a newly introduced or reconfigured manufacturing system to reach sustainable, long-term levels of production in terms of throughput and part quality, considering the impact of equipment and labor on productivity.

Since the RMS paradigm calls for a high frequency of system changes to accommodate production of new products, rapid ramp-up of a manufacturing system after installation and especially after reconfiguration is essential to the success of RMS. Systematic methodologies for root-cause analysis of part-quality problems combined with rapid methods for on-line part inspection are the key for a rapid ramp-up.

Achieving the objective of ramp-up time reduction requires diagnostics and ramp-up methodologies at both the system and machine levels. As production systems are made more reconfigurable, and their functionality and layouts are modified more frequently (see Fig. 19.1), it becomes essential to rapidly tune the newly reconfigured system so that it produces quality parts. If ramp-up is not done quickly, the reconfiguration advantage is lost.

The reason for the possibility of getting deteriorated quality parts after reconfiguration is that reconfigurable machines are mostly designed on the basis of modularized systems, where each of the individual modules has its own interfaces [24]. Therefore, it is imperative to perform diagnostics of the reconfigured system after assembly, and perform the subsequent error calibration and compensation of the system. Consequently, reconfigurable systems must be designed with product-quality measurement systems as an integral part.

The basic engineering steps required are summarized in Fig. 19.18. The measurement step requires the selection of type and configuration of sensor modules (e.g., part dimensions, axis position, cutting force). The diagnostics step utilizes that sensor information to identify errors and faults (e.g., machine or fixture geometric errors, tool breakage). Diagnostics should be embedded at the component level and propagate the information through the machine level to the cell and factory computers. The measurement system and the diagnostic methodology should allow for machine/system diagnosability – identifying the sources for a fault or a part-quality problem. Ideally, the sole cause or source for errors should be identified. The compensation step, either automatically or through operator intervention, en-

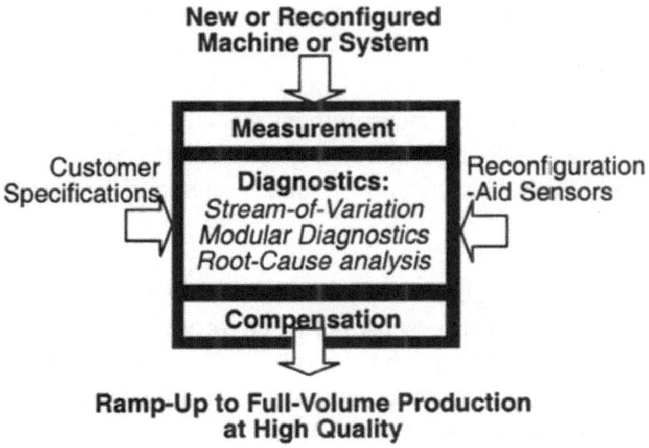

Fig. 19.18 Ramp-up process of RMS.

ables corrective action to be taken (e.g., calibration, adjustment of operating parameters, maintenance).

In order to guarantee short system ramp-up after each reconfiguration, it is necessary to measure product features or process variables that must be selected step-by-step on the basis of the specific problem and situation at hand. To reach this goal it is necessary to have both reconfigurable sensors and inspection devices, and appropriate quality control and reliability tools that help in defining where inspection points must be introduced and which sampling strategy should be applied.

The measurement systems of RMS are intended to help rapidly identify the sources of product-quality problems in the production system and to correct them by utilizing modern information technologies, statistics, and signal-processing techniques. Because of the complex structures of reconfigurable systems, many conventional techniques, such as laser interferometry for the determination of tolerances, are no longer suitable. Consequently, research is currently being undertaken in relation to the suitability of 3-D measurement systems that can measure part and product dimensions in-process [47]. These systems can be subdivided into mechanical (e.g., triangulation or articulated arm with a probe), optical (e.g., interferometry), and ultrasonic systems.

The comparative tests using the different 3-D measurement systems relative to the results from the coordinate-measuring machine yielded results having significant deviations. Thus the practical application of commercial systems has to date been technically unsatisfactory and, on account of the high investment costs, renders the systems non-economical. Within the framework of new research and development work, improved methods for the calibration and error compensation are being sought. In Germany, close co-operation between industry and the universities is taking place within the context of the ACCOMAT Project, as depicted in Fig. 19.19 [48]. The principle of operation of the measurement system is based on a technique well known in navigational engineering and utilizes an intelligent-tooling system in which the sensor and pictuation elements are integrated for its calibration.

Example of System Ramp-Up. Figure 19.20 shows optical sensors on an automotive-body assembly line. These sensors, when properly designed and located, can help diagnose problems on the assembly line (e.g., broken locator, incorrectly programmed robot) that can lead to consumer problems such as wind noise, water leakage, etc. Figure 19.21 shows, with results from actual production, the benefits that can be achieved in terms of rapid reduction of the variation (6σ) in critical body dimensions by applying the methodology depicted in Fig. 19.18 with the Stream-of-Variations (SoV) theory [49]. Note that both the reduction in variation is rapid, and the final level of variation is low.

The SoV methodology combines engineering process models with statistical analysis to account for how product dimensional variations accumulate as the product moves through a manufacturing system. It can be used, with appropriately selected and placed sensors, to diagnose the root causes of the dimensional errors in the production system. This is schematically illustrated in Fig. 19.22 for an automotive-body assembly process [49]:

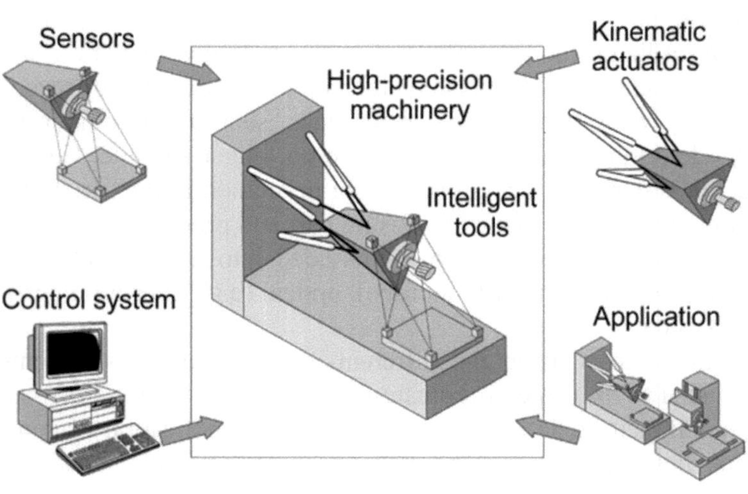

Fig. 19.19 ACCOMAT system for 3-D measurement.

Fig. 19.20 Optical measurement of automotive body dimensions using laser triangulation sensors to achieve detection and isolation of faults.

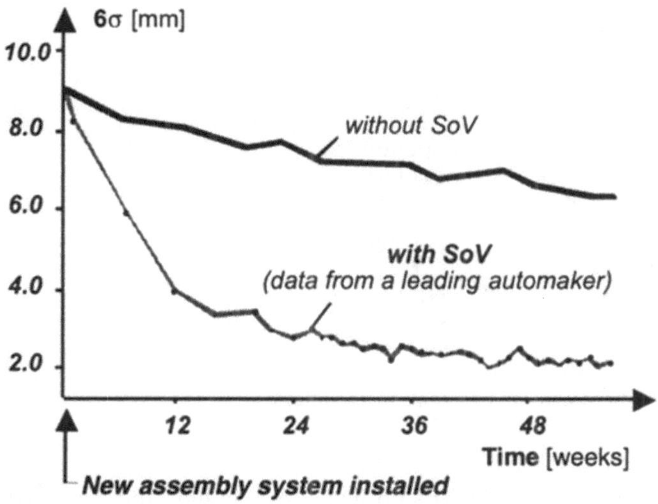

Fig. 19.21 Results showing ramp-up time reduction in automotive assembly.

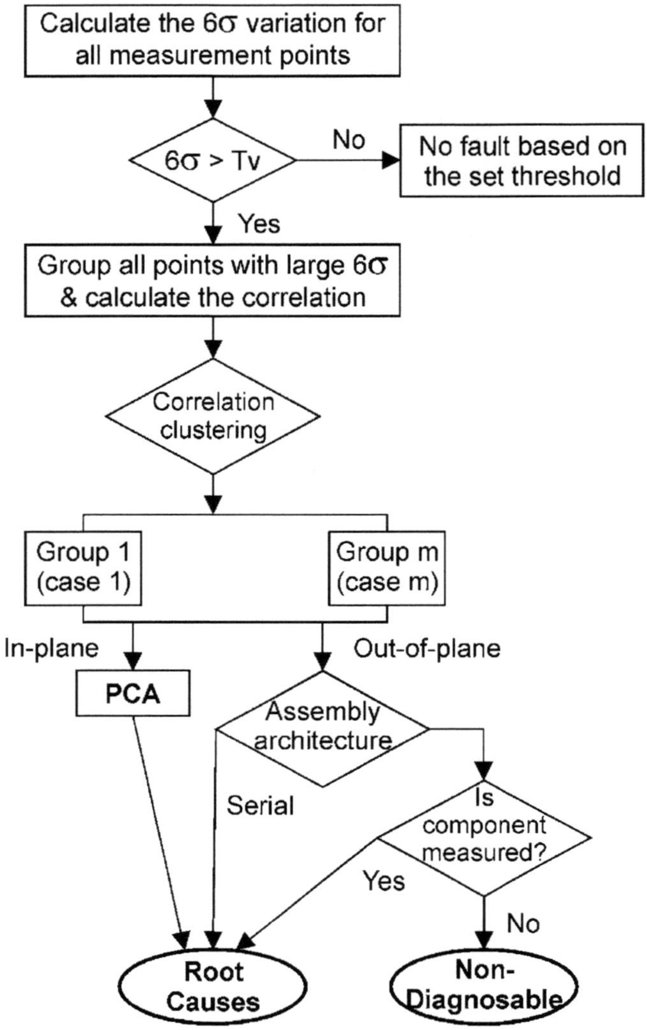

Fig. 19.22 A flow-chart of the stream-of-variations methodology.

- Obtain measurement data for a set of bodies and calculate the standard deviations of all measurement points.
- Group all the measurement points according to the six-sigma threshold, Tv (e.g., 2.0 mm).
- Calculate the correlation matrix for the group of measurement points with six sigma larger than Tv.
- Use clustering analysis to divide the measurement points into sub-groups. Each group forms a case study (i.e., potential problem). The lo-

cation of the process fault is in the assembly station that is one level higher in the hierarchy and has the above-grouped component as input.
- Perform principal component analysis for grouped measurements that have variation in the in-plane direction. Display eigenvectors to animate variation patterns for purposes of visualization. If the grouped measurements have variations in the out-of-plane direction, identify assembly architecture (i.e., serial vs. parallel). Serial assembly is fully diagnosable, while parallel assembly requires additional measurement to be diagnosable.

Using this methodology, fault stations and root causes of variation can be systematically identified. This strategy has been applied to reduce variation, and to shorten launch time, in many automotive-assembly plants. A typical result, from actual automotive-body production, has already been illustrated in Fig. 19.21.

19.9 Conclusions

Global economic competition and rapid social and technological changes have forced manufacturers to face a new economic objective: *manufacturing responsiveness* (i.e., adaptation of the manufacturing system to market conditions). To respond to these challenges, a new type of manufacturing system, a Reconfigurable Manufacturing System, is needed. RMSs are quite different than the current manufacturing technologies (i.e., dedicated manufacturing lines and flexible manufacturing systems) in that the former are designed at the outset with adjustable resources in order to provide exactly the capacity and functionality that are needed, exactly when needed.

Design for reconfigurability is recently emerging as an important new trend in designing manufacturing systems. Therefore, reconfigurable manufacturing is a broad area that continues to attract multi-disciplinary talents for its development. Some of its research topics include:

- Life-cycle economic modeling of manufacturing systems, to determine the product-demand range in which RMS is the optimal choice.
- Automatic generation of system configurations and their impact on throughput, part quality and cost that leads to the establishment of configuration theory.
- Optimal scalability of configurations for demand scenarios.
- Ramp-up methodology for changing configurations at both the machine level and the system level.

- Reconfiguration of monitoring and information systems in an RMS environment.
- Reconfigurable controls for large systems in an open-architecture environment.
- Methodology to generate reconfigurable machines for given manufacturing tasks.
- Relationships between machine builders and users of reconfigurable systems.
- Benefits of reconfigurable systems to society and their impact on job creation.

The industrialized world is faced with the challenge of maintaining its high quality of life and welfare despite global economic competition. A competitive manufacturing industry is required to keep jobs and generate new ones directly, and, through induced services indirectly. To contribute to sustainability, the manufacturing industry must move towards a closed and optimized life cycle of products and related processes [50]. This, in turn, should evolve following the change in economical, social, and technological context, thus achieving sustainability [51]. RMS, hence, may be seen as a methodology for achieving process sustainability at two different levels: (1) tactical level, to respond to turbulence of the demand, internal, and external change, and (2) strategic level, to account for market evolution.

Competition was indeed mainly based on the ability to optimize the technological solution for a given context. In recent years, however, the context has become more and more turbulent: society has new needs in rapid evolution, economy is deeply affected by the evolution toward globalization, and technology exhibits exponential growth. Given such turbulence, competition is more and more based on the ability to quickly adapt to change.

Therefore the emphasis is no longer placed on the optimization of the technological solution in a static way, but on the dynamic problem of continuously adapting the technological solution at low cost. Reconfigurable manufacturing systems can represent one of the cornerstones of this strategy.

Acknowledgements

The authors are pleased to acknowledge the financial support of the National Science Foundation (Grant # EEC-9529125), and the assistance from the many members (from both industry and academia) of the Engineering Research Center for Reconfigurable Machining Systems who provided feedback on the manuscript. The invaluable contributions of Professor T.

Tolio from Politecnico di Milano and Dr. L. Molinari from ITIA-CNR in Milano are gratefully acknowledged. The important suggestions of Dr. Zbigniew Pasek and his assistance in editing the paper are greatly appreciated.

Literature

[1] Hardt D et al (1997) Next-generation manufacturing (NGM) project. Agility Forum and Leaders for Manufacturing. Bethlehem, PA

[2] Koren Y, Ulsoy AG (1997) Reconfigurable manufacturing systems. Engineering Research Center for Reconfigurable Machining Systems (ERC/RMS) Report #1, The University of Michigan, Ann Arbor

[3] Matta M, Tolio T (1999) Flexible and productive manufacturing system architecture. The 15th Int. Conf. on Production Research, Limerick, Ireland, August 1999

[4] Ulsoy AG, Heytler P (1998) A survey of flexible and reconfigurable manufacturing systems, Report # 11. ERC/RMS, The University of Michigan, Ann Arbor

[5] Koren Y et al (1998) (Eds) Open architecture control systems: Summary of global activity. ITIA Series, vol 2. Milano, Italy

[6] Mehrabi MG, Ulsoy AG (1997) State-of-the-art in reconfigurable manufacturing systems, Report 2, vol. I and II. ERC/RMS, The University of Michigan, Ann Arbor

[7] Aronson RB (1997) Operation plug-and-play is on the way. Manufacturing Engineering, pp 108–112

[8] AMT (Association for Manufacturing Technology) report (1996) A technology road map for the machine-tool industry.

[9] Ashley S (1997) Manufacturing firms face the future. Mechanical Engineering, pp 70–74

[10] Garro O, Martin P (1993) Towards new architecture of machine tools, Int. J. Prod. Research 31 (10):2403–2414

[11] Rogers GG, Bottaci L (1997) Modular production systems: A new manufacturing paradigm. J Intel Mfg 8:147–156

[12] Koren Y (1983) Computer control of manufacturing systems, Chapter 10, McGraw-Hill, New York

[13] Atkins WS (1990) Management consultants. Strategic Study on the EU Machine Tool Sector

[14] Heisel U, Michaelis M (1998) Rekonfigurierbare werkzeugmaschinen. ZwF Zeitschrift für wirtschaftlichen Fabrikbetrieb, Vol. 93, No. 10:506–507

[15] Pritschow Get al (1993) Open system controllers – A challenge for the future of the machine tool industry. Annals of the CIRP 42 (1):449–452

[16] Van Brussel H, Braembussche PV (1998) Robust control of feed drives with linear motors. Annals of the CIRP 47 (1):325–328

[17] Koren Y, Kota S (1999) Reconfigurable machine tools, U.S. Patent 5 943 750

[18] Paredis C et al (1996) A rapidly deployable manipulator system. Proc. IEEE ICRA, pp 1434–1439

[19] Fukuda T, Ueyama T (1993) Autonomous behavior and control. Proc. IEEE ICRA.

[20] IMS Report (1990) Joint international research programs into an intelligent manufacturing system. International Robotics and Factory Automation Centre (IROFA)

[21] Bollinger J et al (1998) Visionary manufacturing challenges for 2020. National Research Council Report, National Academy Press, Washington

[22] Lee GH (1997) Reconfigurability considerations in the design of components and manufacturing systems. Intl J AMT 13 (5):376–386

[23] Thielemans H et al (1996) HEDRA: Heterogeneous distributed real-time architecture. Control Engineering Practice 4 (2):187–193
[24] Heisel U et al (1998) Auslegung von maschinenstrukturen mit gelenkstab-kinematik. wt Werkstattstechnik, Vol. 88, Nr 4:183–186
[25] Mehrabi MG et al (1998) Future trends in manufacturing: A survey of flexible and reconfigurable manufacturing systems, submitted to J Mfg Sys
[26] Eversheim W (1989) Organisation in der produktionstechnik, Band 4: Fertigung und Montage. VDI-Verlag, Düsseldorf
[27] Sethi AK, Sethi SP (1990) Flexibility in manufacturing: a survey. Intl J FMS 2:289–328
[28] Grieco A et al (1995) Simulation of tool and part flow in FMSs. Intl J Prod Rsch, 33 (3):643–658
[29] Chryssolouris G (1996) Flexibility and its measurement. Annals of the CIRP 45 (2):581–587
[30] Koren Y et al (1998) Impact of manufacturing system configuration on performance. Annals of the CIRP 47 (1):369–372
[31] Tolio T, Matta A (1998) A method for performance evaluation of automated flow lines. Annals of the CIRP 47 (1):373–376
[32] Nèmeth I et al (1999) Conceptual design of 3-axis machine tools. Proc. 32nd CIRP Intl Sem Mfg Sys, Leuven, pp. 239–248
[33] Tönshoff HK et al (1994) Intelligence in machine tools by configuration. 7th Intl Conf Prod/Precision Eng, 4th Intl Conf High Tech, September 1994, Chiba, Japan
[34] Erixon G (1996) Modularity – the basis for product and factory re-engineering. Annals of the CIRP 45 (1):1–4
[35] Pasek ZJ et al (1999) OpenFront: system-level software for configuration and control of manufacturing systems. ERC/RMS, Report #21, The University of Michigan, Ann Arbor
[36] Shamoto E, Moriwaki T (1997) Rigid XY table for ultraprecision machine tool driven by means of waling drive. Annals of the CIRP 46 (1):301–304
[37] Heisel, U et al (1995) Forderung an hochgeschwindigkeitsmaschinen. wt-Produktion und Management, No. 85:155–161
[38] Pritschow G et al (1998) Research and development in the field of parallel kinematics systems in Europe. First European-American Forum on Parallel Kinematics Machines: Theoretical Aspects and Industrial Requirements, September 1998, Milano, Italy
[39] Heisel, U (1998) Precision requirements of hexapod machines and investigation results. First European-American Forum on Parallel Kinematic Machines, September 1998, Milano, Italy
[40] Weck M et al (1998) Parallel kinematics – the importance of enabling technologies. First European-American Forum on Parallel Kinematic Machines, September 1998, Milano, Italy
[41] Kieser D (1997) In der werkzeugmaschinentechnik bricht eine neue aera an. Industrie Anzeiger, Vol. 22:38–41
[42] Kota S (1999) Design of reconfigurable machine tools. Proc. 32nd CIRP Intl Sem Mfg Sys, May 1999, Leuven, Belgium, pp 297–303
[43] Pritschow G et.al. (1997) Modular system platform for open control systems. Production Engineering 4 (2)
[44] Lutz R et.al (1997) Konfigurierungswerkzeuge für offene steuerungen. Innovation durch Technik und Organisation. Berlin, Springer
[45] Koren Y et al (1996) Real-time open control architectures and system performance. Annals of the CIRP 45 (1):377–380

[46] Sperling W et al (1996) OSACA ESPRIT 6379/9115 final report. FISW GmbH, Stuttgart, Germany
[47] HÜMNOS (1998) Abschlußbericht. VDW Forschungsberichte, Frankfurt
[48] Spath D, Mussa S (1999) ACCOMAT - Die genauigkeitsgeregelte maschine. wt Werkstattstechnik, Vol. 89, No. 5:235–238
[49] Hu SJ (1997) Stream of variation theory for automotive body assembly. Annals of the CIRP 46 (1):1–4
[50] Boer CR, Jovane F (1996) Towards a new model of sustainable production: ManuFuturing. Annals of the CIRP 45 (1):415–420
[51] Jovane F (1998) Concepts for strategic machinery innovation, Initial Meeting of ARMMS Networks, Bruselles, December 1998

Chapter 20
Robot Technology

R. D. Schraft

20.1
Introduction

The demand for high-quality automation solutions remains constant. Existing deficits and rapid advances in innovations, especially in the fields of information and sensor technology, aim to improve the abilities of robots with regard to their performance to adapt to changing manufacturing conditions, their ability to be quickly configured to carry out production tasks, and their intuitive operation in all operating modes from installation and operating tasks right up to maintenance. Subsequently, many new areas of use are being opened up for robot systems in production and service. In many cases, the intelligence of an automation system for performing tasks is still insufficient. As well as further developing intelligent system behavior, the path is increasingly being followed whereby the intelligence and skill of the operator is used to perform tasks cooperatively. The objective here is to develop systems, components, and processes for communicating and interacting with assistance systems which offer the user the possibility not only of direct control but also of having mechanical sub-tasks carried out independently ("laborer"). The sensory as well as the actuatory forming of the assistant allow for the development of completely new forms of interaction in complex or complexly modeled surroundings.

The "growth rate" of German mechanical engineering was 0% in 1999 [1]. In the four disciplines of robotics and automation (R+A), i.e., robotics/handling, assembly, industrial image-processing, and systems and components, the situation looks much better: For the fourth time in a row, an increase in turnover of 21% to about $4 billon US or 8.5 billion German marks proves that there is a mounting demand for high-quality automation solutions, mainly motivated by economies in production costs (Fig. 20.1) [2].

From the user's point of view, technical deficits arise in four areas in automation technology, which result in a demand for standardized, modular system architecture and the use of higher levels of technical intelligence (Table 20.1).

Consequently, the deficits mentioned below place demands on future automation solutions, which place an emphasis on the use of sensors and new-style forms of information technology.

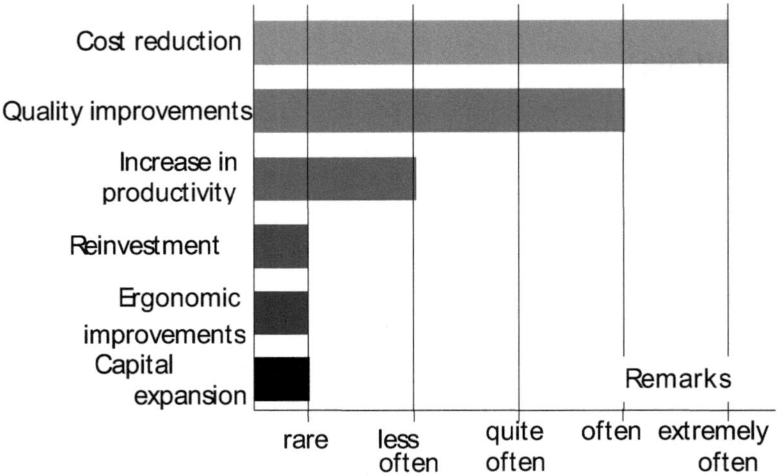

Fig. 20.1 Grounds for deciding to automate production processes [2].

Table 20.1 Current deficits in automation technology [2].

Operation	- Only intuitive procedures possible for the user.
Programming	- Insufficient software re-use and flexibility. - Inadequate standardization/ modularization of control software and hardware.
Abilities of sensors	- Inadequate performance of sensors (sensitivity, signal editing, etc.). - Insufficient abilities of sensors to be integrated into subsystems/ equipment. - Inadequate functional certainty of sensors (especially under difficult conditions).
Interfaces	- Insufficient conformity between manufacturers (either between differing manufacturers or even differing operating materials made by the same manufacturer).
Flexibility	- Increasing amount of variants and smaller lot sizes make it very difficult for operating materials to be adjusted („flexibility“) to suit varying product parameters. - High demands of work piece features (particularly large or small, bending flaccidity, etc.) still make automation difficult.

These deficits and the rapid rate of innovational progress, especially in the fields of information and sensor technology, aim at promoting robots to become central technical intelligence components within automation systems.

System characteristics which, in conjunction with modern information and sensor technology, can be realized in automation technology as technical intelligence are:

- **Adaptability**: the ability to carry out changes with regard to workpiece variants, site, the arrangement of workpieces, and environmental conditions.
- **Changeability**: the rapid reconfiguration of equipment software and hardware to meet changing task requirements ("Rapid Deployment Automation," RDA).
- **Autonomy**: Robust, automatic task execution even in fringe situations.
- **Man-machine interaction**: Intuitive and interactive operation in all operating conditions (set-up, operation, maintenance).

The trend towards using intelligent components in standardized system architectures which can be operated by all users is in full spate. With this as a backdrop, the role of industrial robots (IR) in industrial automation is becoming especially important.

The following paragraphs describe current trends in the development and implementation of robot systems using modern sensor and information technology. The main themes of the central functions of technical intelligence such as adaptability, changeability, and man-machine interaction are presented here.

20.2 Present Situation of Robot Utilization and Robot Technology

The dynamic development of R+A in Germany has mainly been due to the structure and teamwork of component manufacturers, system integrators, and users [3]:

- The automotive industry still makes up almost half of all new IR installations.
- Technologically perfected applications, such as spot- and line-welding, are particularly economical and make up approximately 40% of all new installations.
- A high number of specialized small- and mid-sized-system companies cover niches in IR applications. Here, most IR's are externally bought components which make up approximately 30% of the average price of a system (engineering, assembly, and initial operation: 30%; components: 30%).

- In comparison with wage-piece costs, component costs are falling all the time. As a result, the price of an IR fell by almost 30% between 1990 and 1997. Simultaneously, the performance of these devices rose considerably, especially with regard to driving speeds (Fig. 20.2).
- The use of material automation (palletizing, commissioning, and handling) is on the way to increasing applications enormously in the chemical and pharmaceutical industries and in the mail order trade.

The speed of innovations in R+A is becoming increasingly influenced by mass-market technologies (see Fig. 20.3). Large piece numbers justify costly developments in hardware (processors, periphery, sensors, etc.) and

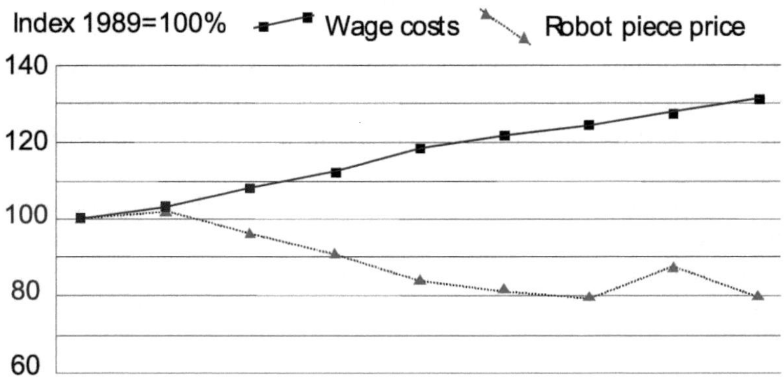

Fig. 20.2 Indexed wage costs compared with the average piece price of a robot.

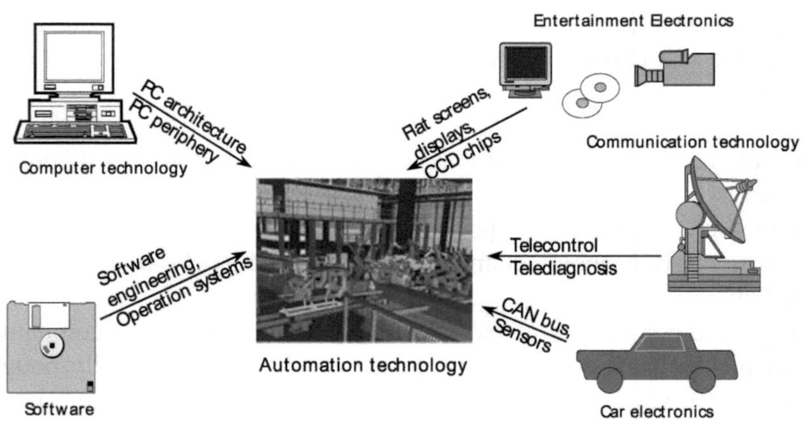

Fig. 20.3 Diffusion of mass-market technologies into automation technology.

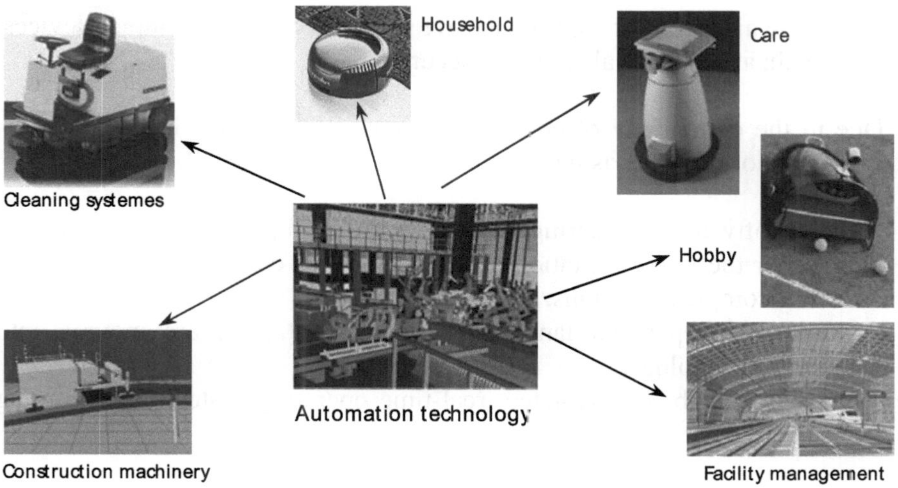

Fig. 20.4 Diffusion of automation technologies into new fields of application.

in software (development platforms, operating systems, visualization tools, etc.). Because of the wide distribution, defacto standards associated with system architecture and interfaces (e.g., TCP/IP, CAN, USB) are being created, which simplify component integration.

In the same way, a diffusion of R+A technologies from industry into such new fields of application as the construction industry, medicine and rehabilitation, etc., can be observed (see Fig. 20.4).

20.2.1 Trends in Control Technology

The PC technology breakthrough into control technology has taken place. As a result, valuable intelligent automation potentials have been opened up [4]:

- A dramatic improvement in the cost-performance ratio of control, communications, and visualization hardware.
- The support of open, modular system architectures for flexible, customized solutions.
- The extensive availability and easy inclusion of software modules as "components" (also from other manufacturers) such as sensor data-processing, soft SPS, process visualization, simulation, and diagnosis.
- The creation of a defacto standard for software engineering (methods and tools): complete integration into data networks (field bus, TCP/IP, USB) from process level right up to company management level (ERP).

- The availability of comprehensive periphery such as input devices, visualization material, and data security.

Due to the complexity of control functions, demands for installable software technologies are rising:

- New software engineering methods such as object-orientated development or the implementation of agent concepts contribute towards mastering complex functions.
- Platform-independent languages, especially JAVA, are moving into control technology.
- The use of object-orientated, real-time operating systems is becoming standard.

The developments and applications of modern methods in control technology and software engineering are shown using the following three examples.

20.2.1.1
Intelligent Cross-Section Components IQ2000

In Project IQ2000, together with control manufacturers, decentralized sensor and control modules are being developed which are connected via bus systems (CAN bus). The control software for user-specific developments is based on the comprehensive real-time library. Technological functions for movement control and sensor data-processing are available for a wide range of applications, especially for mobile-handling robots. Figures 20.5 and 20.6 show the architecture and development process for advances in control used in IQ2000.

20.2.1.2
Holonic Manufacturing Systems (HMS)

Holonic controls represent distributed intelligent systems based on multiagent systems. In 1997, CORBA middleware was standardized as a communications platform for agents. Holonic controls are still in the development stage, and control concepts are currently being tested on demonstration equipment. The emphasis is on the use of holonic concepts for the automated control of material flows (e.g., using mobile robots) in production environments.

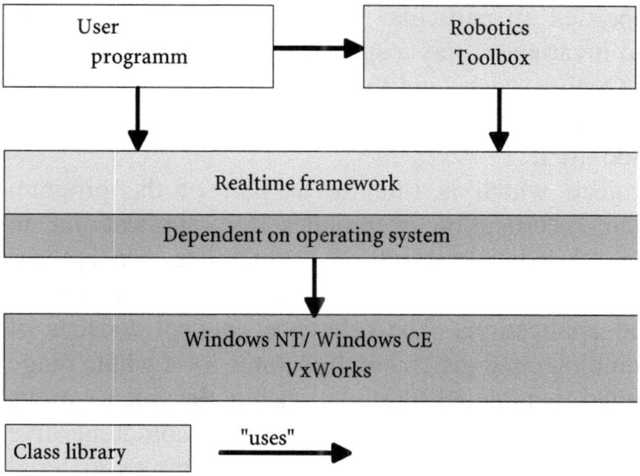

Fig. 20.5 General structure of the class library used in the IQ2000 concept.

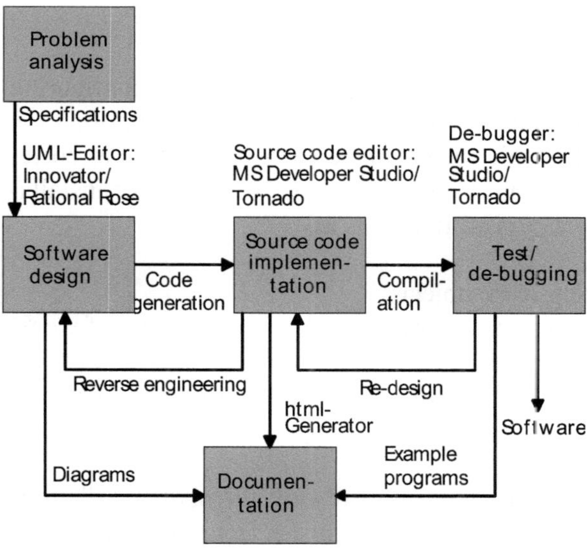

Fig. 20.6 IQ2000: Software development process and aids used.

20.2.1.3
Software Specifications for Automation Systems

The speed and quality of software developments in R+A are going to be the success factors in increasingly complex systems. The percentage of software development costs already make up 75% to 80% of engineering and integration costs. The specifications of the system according to system characteristics and properties, in particular with regard to functionality, behavior, structures, and invariance, play a special role in the development of automation systems. Quality, costs, and the speed of implementation of automation software are directly dependent upon the type and user-friendliness of the method used.

The aim of the project, which is running as part of the program "Integration of software-specification techniques for applications in engineering science," is the development of a reference concept for developing software-intensive technical systems, especially in the field of robot developments and applications. The reference concept consists of creating a standard terminology and general architectures for a whole range of problems. In a so-called metamodel, the relationships between various specification techniques can be expressed in a uniform and comprehensive model. The demonstration is the automated control of a complex material (e.g., using mobile robots) in production environments.

20.2.2
Trends in Sensors

Selecting suitable sensors for carrying out the tasks required, configuring the sensor system, and processing sensor data into movement information are costly engineering tasks which need to be carried out for the corresponding automation task. Up until now, less than 10% of robot installations possess sensors for such typical functions as:

- Localizing and identifying work objects.
- Process guidance using contour-following or the dimensioning of working or jointing forces or moments.
- Environmental recognition, such as the active safeguarding of an area or collision detection.

At present, image-processing (IP) is used in 6% of all robot installations [4]. Most have PC-based standard sensor systems, a fact which simplifies their integration into IR controls and will contribute to their rapid spread in R+A. With rising piece numbers (presently a rate of increase of over 25%/year), a fall in price of systems while simultaneously improving performance is being observed [5].

Two examples demonstrating current activities with regard to the development and implementation of sensor systems are given below:

20.2.2.1
Sensor Toolbox

In a similar way to Project IQ2000, a sensor toolbox aims to support the configuration and implementation of sensor systems by using and easily including object-orientated software components. For this, modules are available as object classes for various sensors such as laser scanners, ultrasound sensors and inertial-measuring location sensors (acceleration sensors, inclinometers and gyroscopes).

20.2.2.2
"Bin Picking" Using Tire-Sorting as an Example

On the one hand, the task of order of arrangement enables the removal of feeding, thinning out, and ordering equipment, but on the other hand, the position and identity of the parts need to be recognized every single second by sensors.

With tire-sorting, this is carried out effectively by a 3-D laser scanner (Fig. 20.7):

- A 3-D laser scanner records the scenario.
- The modeled tire geometries are matched up to give the position of the tires.
- Collision-free gripping points and movement paths are calculated for the IR and handed over to the control system.

Fig. 20.7 Automatic sorting of unsorted tires: 3-D laser scan of the scenario (left), original scenario (right).

20.3 Changeable and Adaptive Robot Systems

As the technical intelligence of automation systems will be augmented in the future through the use of IR, fewer specific periphery elements in the form of feeding, thinning out, and supply equipment will generally be necessary.

With open, modular system architectures, the general system configuration will become more modularized to further reduce the amount of expensive integration and initial operation work required. In this way, new operator models of automated solutions can be expected, which make systems available to the production operation on a time-limited basis as a production service, and which are then used to perform other tasks.

The flexibility and adaptability of automation systems may also have strong repercussions on a company with regard to product strategies in the form of mass customization. The technical requirements for customized product configuration, flexible work-planning, and production still need to be created to a large extent and are the object of intensive development efforts.

20.3.1 Man-Machine Interaction

The limits of automation are reached when task execution requires sensory capabilities, knowledge, and skills which can no longer be economically

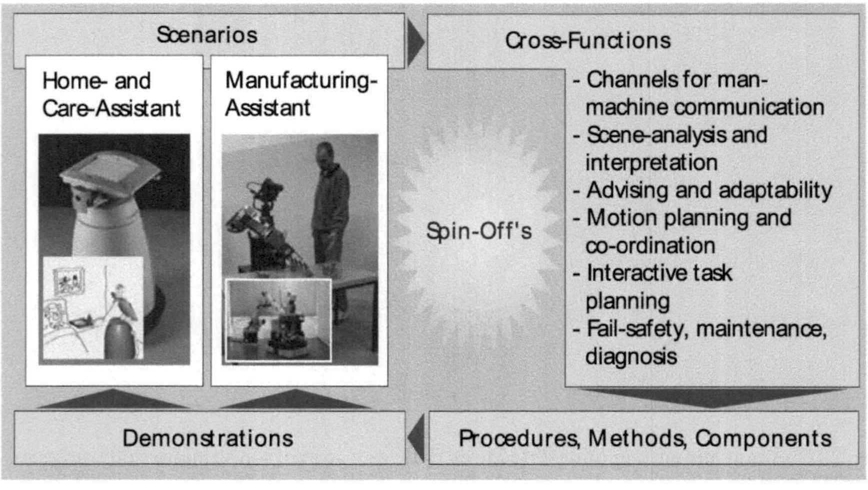

Fig. 20.8 Structure of the master project MORPHA: Scenarios and R&D themes.

performed using machines. However, some simple or burdensome subtasks may still be performed by machine as part of task execution. This form of work-sharing could be flexible and include various modes of operation of the machines, from completely interactive right up to completely autonomous task execution. As an overall result, the objective would be to have economical, cooperative, and flexible task execution. [6]

In the BMBF (German Federal Ministry of Research) master project MORPHA, key questions are being worked on associated with the flexible physical and informatory cooperation between man and machine (Fig. 20.8).

The two following examples aim to illustrate cooperative task-sharing between man and machine: a production assistant, and assistant systems in the home and in the field of nursing.

20.3.1.1
Scenario 1: The Production Assistant

A new quality in improving production processes with regard to increasing productivity and humanizing places of work could be achieved by using so-called production assistants which perform tasks interacting with humans, where man assumes the command, monitoring, and teaching functions. In situations where the machine "doesn't know," man intervenes to guide and further instruct and is a partner in the work-sharing production process.

In future production scenarios, the production assistant will support man, for example, in carrying out handling, transport, and inspection tasks in a typically complex production environment with machine tools, conveyor belts, storage areas, etc. Typical tasks are the gripping of unsorted parts, transportation to a machine tool or assembly area, assisting in assembly or transportation to an inspection or measuring area, etc.

In order to be able to manage such tasks in a factory environment often subject to change, a high level of flexibility is necessary. These changes could include transport crates having variable positions or workpieces having different forms and weights in unsorted positions in crates requiring transportation to various machine tools or target objects.

20.3.1.2
Scenario 2: Assistant Systems in the Home and in Nursing

Visions assume that assistant systems will be used as mass products in our every-day lives (in the home) for a whole variety of tasks which crop up. These assistant systems comprising multi-media interfaces, handling arms, and grippers, satisfy all the demands of a technologically and economically sophisticated mass product with a substantial mid- to long-term market potential. The home or nursing assistant should enable the user, despite any handicaps he/she may have, to carry out a whole range of simple tasks without requiring external help. A home or nursing assistant is typically capable of the following functions:

Personal care by performing simple tasks:

- Bringing objects.
- Helping to grip, lift, and hold objects and devices.

Mobility assistance:

- Guiding and supporting, e.g., helping to get out of bed or stand up from a chair, as standing or walking assistance, or helping to reach the bathroom.
- Transporting a person using either a self-driving wheelchair or buggy.

Cleaning:

- In a single step, the assistant is taught how to clean objects and surfaces in its environment such as floors, wash basins, or walls.
- Communications and domestic technique management (by being linked up to information networks).

Similar to the production assistant, a high level of flexibility with a high level of user-comfort and safety is required to perform a wide variety of tasks in sometimes unknown surroundings.

The most important, common features of assistance systems applying to all scenarios are the result of the type of surroundings in which the person and the machine move around and the natural working and living environment of the person, and of the requirements of the interaction between the person and an autonomously or partially autonomously moving and acting machine which arise in a similar form in both scenarios.

- In both of the scenarios described, it is presumed that the environment is unprepared, moderately dynamic, unstructured, and is full of moving people.
- Man and machine cooperate so that, in the context of specific tasks, abilities may be used in an optimum way.
- As far as movement, handling, and gripping tasks are concerned, the work-sharing interaction between man and machine is similar in both

scenarios. A differentiation is made between interaction with and without physical contact between the assistant and the person:

- Interaction without physical assistance
  - Controlled approach, passing objects, e.g. tools or assembly parts in the case of the production assistant, and books, meals, etc. in the case of the home or nursing assistant
  - Running errands
  - Localizing and gripping objects
  - Performing processing or maintenance jobs
- Interaction with physical contact
  - Interactive handling and shared movement, e.g. lifting loads (production assistant) or guiding people (home and nursing assistant)

The two scenarios described are representative for a wide range of assistant systems (Table 20.2).

The fundamental mechanisms for cooperative task execution are generally independent of the type of application and are grouped into functional requirements. These are being developed in the MORPHA project and have been divided into five sections:

- Man and machine must be able to clearly understand one another even in complex situations. The exchange of information must be intuitive and comprehensible, and use several sense channels. These comprise mainly speech, haptics, and graphics. This leads to the section concerned with man-machine communication channels.
- A prerequisite for efficient cooperation is the sensorial recording of the environment by the assistant and an understanding of the task set and its context. This leads to the next section concerned with scene-analysis and interpretation.
- Effective assistance does not only require a machine's own intelligence but also a transfer of knowledge between man and machine and a certain ability of the machine to learn and adapt. This leads into the next section concerned with teaching and adaptability.
- The flexible work-sharing task execution between man and the assistance system is based on movement and handling sequences. These must be planned, coordinated, and, where necessary, be mutually adapted. This leads to the section concerned with movement-planning and coordination.
- With systems which physically cooperate with man, high demands are placed on functional and operational safety. Complex systems also need

Table 20.2 Conceivable assistant systems with various applications.

System	Brief description
Medicine	
Surgeon simulator	Simulation of surgical operations during professional training or for pre-operative practice with multi-media support for the surgeon supplied by the training station
Rehabilitation	
Partially autonomous wheelchair	Wheelchair is operated by the handicapped person. It can be driven safely in an environment full of moving people.
Rehabilitation assistant	Therapy with highly-sensitive guidance of parts of the body, training for parts of the body
Building trade	
Assistant for transport and handling in construction work	Mobile platform with multi-axial kinematics for assisting in transport, processing, assembly and handling tasks („helping hand“)
Automatic digger	Excavating in areas difficult to access or hazardous surroundings (retreat)
Trade, transport and traffic	
Courier service in large-scale buildings	Delivery of post and materials with handovers in large-scale buildings such as airports, railway stations, hotels and office buildings
Intelligent sales system	Handling assistance for registering and packing goods
Security, radiation and catastrophe protection	
Fire-fighting	(Partially) autonomous assistant for fire-fighting, rescue or surveillance
Handling tasks in contaminated environments	Assistant system for general tasks in contaminated environments, de-fusing bombs, clearing mines, etc.
Home, hobby, leisure	
Museum guide	Robot guides people in public environments, addresses people, gives information on request, is an attraction
People mover	Platform with seating or standing room, luggage rack, for transport within spacious buildings

to be easily operated and maintained. This leads to the final section on safety, maintenance, and diagnosis.

It should be possible to put some of the results into practice for products before the project has even been completed. First, these aim at improving man-machine interaction, starting with new programming and operating consoles, person-recognition systems and even intelligent, human sense-orientated control for mechatronic systems. Secondly, the development and supply of „anthropomorphic," intuitive operator and program interfaces, should pave the way for new-style assistant systems in productions, in the home, for nursing or even for medicine.

Literature

[1] Müller S (1999) Robotik + Automation im Zentrum der Investitionsgüterindustrie – Ssandortexzellenz durch Technologiekompetenz Press Conference Hanover Fair 19 April 1999

[2] Schraft RD, Kaun R (1999) Automatisierung: Stand der Technik, Defizite und Trends in der Automatisierungstechnik. Verlagsgruppe Handelsblatt GmbH, Wirtschaftswoche, Düsseldorf

[3] Karlsson J (1998) World robotics 1998. United Nations UN/ECE

[4] Atherton RW (1998) Moving JAVA to the factory. In: IEEE Spectrum, December-Vol. 35 (1998) No. 12, p.18–23

[5] Ersü E (1999) Bildverarbeitung ist der Motor für den Mechatronikmarkt – Wer führt wen? Vision den Roboter oder umgekehrt Press Conference Hanover Fair 19 April 1999

[6] Schraft RD, Schmierer G (1998) Serviceroboter – Produkte, Szenarien, Visionen. Springer, Berlin Heidelberg New York

Chapter 21
Methods for Nondestructive Testing and Diagnostics of Automatic Equipment and Technological Systems of Machines

V.V. Klyuev, E.G. Nachapetjan, V.V. Scherbakov †

21.1
Nondestructive Testing and Diagnostics of Automatic

Equipment

The philosophy of safety and quality management has changed from «to react and correct» to the principle "to predict and reduce the losses".

Quality is important at all levels since it involves technological processes, service, and environment. Great efforts are being put into investigating the relationships between quality, innovation, and research.

More than hundreds of physical methods and thousands of different types of units machine are used today in non-destructive testing (NDT) and diagnostics for approaching technological problems of quality [1–6].

The machine-tools and assembly lines include or use many NDT and diagnostic units to test equipment while it is manufactured, inspected, installed, and repaired to make sure that the tested object complies with its technical requirements and specifications. One of the trends in developing automated, highly informative systems is their intellectualization. The expert systems should have the database about the particular object (qualimetric information), its functioning condition (diagnostic results), variety of types and sizes of defects and faults, and reliability characteristics. For the expert system of the interactive type, it is typical that the information obtained by means of the measuring instruments is initial primary [1–3]. The qualimetric information is useful both for diagnostics and for machine design as the basis for working out new calculations and simulation methods, and for forecasting productivity and reliability [7–10].

The elaboration of practical ways for obtaining qualimetric information under laboratory conditions, in production and operation, and the information concentration in the database are also shown here. Also, the examples of solving practical problems on the basis of using the diagnostics and NDT methods are described.

21.1.1
Purposes and methods of NDT and diagnostics

In Fig. 21.1 ways of using NDT methods and technical diagnostics for solving technogenic, environmental, medical, and criminalistic problems are presented [1].

The main diagnostic purposes and methods are [1–9]:

- Elaboration of quality criterion based on qualimetry of mechanisms;
- Choice of the nomenclature of quality indices;
- Definition of the range for quality indices and parameters of mechanisms, based on their purpose and service conditions;
- Development of the mathematical models of mechanisms;
- Fault analysis of mechanisms (design and production quality);
- Fault and reliability analysis of similar mechanisms during operation (like distribution, failure intensity, and safety analysis);
- Development of methods of vibro-acoustic dynamics and diagnostics;
- Comparison of various test strategies and associated test and fault costs;
- Working-out guides for detection of events, volumes, and periods of diagnostics, tolerances for measured parameters, fault-tree analysis, and structure of integrated indices;

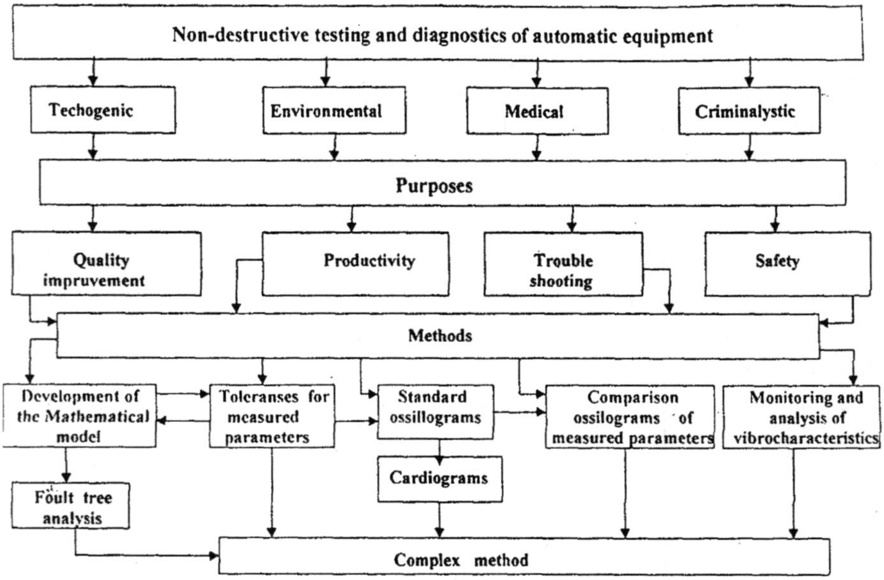

Fig. 21.1 Methods of non-destructive control and diagnostics of automatic equipment.

- Selection of transducers and units for obtaining diagnostic information and of production quality test systems;
- Development of metrology algorithms' library, diagnostic software, and database;
- Personnel training.

21.1.2 NDT Methods

NDT is often used for diagnostics of machines' technological systems. NDT is based on physical processes of interrelation between a physical field and a substance with the tested object. For technological systems of machines, machine parts, and materials many methods are used: electric, magnetic, eddy-current, microwave, electromagnetic, infrared, optical, radiographic, ultrasonic, and penetrant method. For testing technological machines' hydraulic systems, we use leakage testing; for rapid-rotation systems, vibration analysis [1–6, 11].

21.1.2.1 Electric NDT Methods

An electric NDT method creates an electric field in the tested object by a direct action of electric disturbance aimed at that object, or an indirect action of non-electric disturbance (mechanical, etc.). The tested object's electrical characteristics are used as initial informative parameters.

The electric NDT methods can be divided into three groups according to purpose: methods based on measuring the parameters of composition and structure of a material, those based on finding geometrical dimensions and mechanical stresses of a tested object, and those based on finding moisture level.

21.1.2.2 Magnetic NDT Methods

Magnetic NDT methods can detect microdefects, e.g., cracks, blowholes, and incomplete fusion areas. It has been an effective practice to test the condition of surface layers, quality of surface hardening, nitration, and so on.

Ponderomotive testing is based on measuring the force or the moments of force acting in a constant magnetic field, or the force pulling a permanent magnet or electromagnet from the item to be tested, or the same for the moment. Magnetostatic testing is based on measuring magnetic permeability of the tested material. Induction testing is based on measuring of a combined resistance or inductance of the measuring coil, etc. Induction thickness gauges are widely used today to measure thickness of nonmagnetic coatings on a ferromagnetic base. They are based on identifying changes in magnetic resistance of a magnetic circuit.

Magnetic techniques are widely used nowadays for making metal detectors.

21.1.2.3
Eddy-current NDT Methods

For an eddy-current-measuring transducer, induction coils are used. Testing may be conducted without contact of the eddy-current transducer and the tested object at a distance large enough to allow free movement of the transducer with reference to the object.

Flow detectors fitted with surface eddy-current transducers are used for testing rods and pipes (diameter or thickness of metal of pipe walls); they are made of ferromagnetic and low-magnetic types of steel as well as of non-ferrous metals and alloys and are used to determine the degree of mechanical stress, to monitor the quality of surface layers, and so on.

21.1.2.4
Microwave Electromagnetic NDT Methods

This method is based on the analysis of the interaction between the microwave frequency of electromagnetic emissions in the radio frequency range (with a wavelength from 1 to 100 m) and the tested object.

In testing we distinguish: amplitude, phase, frequency-phase, and polarization-geometric and time-based. It is possible to obtain information on the internal condition of the tested object either through its interference picture or its restored image. When using several frequencies – by a shift of the resonant absorption frequency, by way of comparison of two or more frequencies, or by analysis of the spectrum of frequencies – the radiopolarization technique is used for studying residual stresses and stress deformations.

21.1.2.5
Infrared NDT Methods

The infrared NDT methods are based on recording temperature fields of the tested object. In most cases, defect analysis is made by scanning the surface with a narrow optical beam, heat-setting ink, and so on. The local difference of temperatures across the defected and defect-free areas is obtained by scanning.

Today infrared topography – the visualization of internal sections of an object based on thermal effects – is gaining more and more application.

21.1.2.6 Optical NDT Methods

The informative parameters of optical radiation are its space and time distributions in terms of its amplitude, frequency, phase, polarization, and coherence degree.

Defects which can be detected by the optical techniques are: air holes, spills, pores, cracks, foreign object inclusions, internal stresses, structural changes in materials and their physical and chemical properties, deviations from the assigned geometry, etc.

Illumination can be provided by laser, collimator, and by a light diffuser. The image analyzer consists of an optical system and a photodetector. In machine building for control of dimensions, the method of optical projection is used.

To measure dimensions, magnifying glasses, microscopes, television, laser, and raster systems are applied.

By profilographical methods of roughness inspection, the principle of an optical arm is used to measure each movement of the needle that contacts with the tested surface.

The holographic endoscopy appears to be very promising for monitoring defects present in parts of items with different access. The most advanced method is the method of multi-aspect tomography.

The so-called technical vision systems have the widest application. This system includes preliminary processing and image analysis for assortment of tested objects.

21.1.2.7 Radiographic NDT Method

An object is tested with X-ray at different types of ionizing radiation (roentgenography, gammagraphy, based on accelerators and neutron radiography). The simplest method is fluoroscopy. Today film radiography and xeroradiography are distinguished.

21.1.2.8
Ultrasonic NDT Methods

The ultrasonic methods are based on the transmitting and receiving of the elastic waves of the tested object (workpieces and finished products). For diagnostics, piezoelectric, electromagnetoacoustic, laser, and magnetostrictive transducers are used.

The basic methods are: echo, shadow, resonant, velosymmetrical, impedance, free-vibration, and emission (AE) methods.

The most widely used passive method is based on acoustic emission. When testing metal-cutting tools, the waves are produced by the object itself. The echo method is also gaining the greatest practical application. This method is effective for solving flaw-detection problems arising in forged pieces, castings, welds, and many nonmetallic materials.

The research work in Russia and the USA has shown that the radiotelevision testing allows the inspectors to reveal a higher portion of defects, comparing the ultrasonic testing with the following inspection by radiography. This concentration has led to the development of the system that tests joints by the ultrasonic method and the radiotelevision method together.

For ultrasonic methods of testing, the work is continued for further improvement of sensitivity and resolution capacity of instrumentation, increasing the reliability of measurement. The trend still exists for higher ultrasonic frequencies, simultaneous use of multi-transducer units, information presentation on two- and three-dimensional displays, and a more comprehensive mode instead of the simple scanning, and the wide use of digital processing of signals.

Diagnostics of bearings at their manufacture and operation is run with use of acoustic-emission (AE) and vibroacoustic methods. At manufacturing, the materials and parts are influenced by testing load, indentor's force, and laser beam. Vibrations are within a frequency band of 100–500 kHz. In operation, fault analysis of the whole bearing and precision analysis of the surfaces' wear and destruction are made.

21.1.2.9
Penetrant NDT Methods

These methods are based on indicator-fluid penetration into cavities on the surface and through holes in the material of the tested object (making a tracer). NDT methods are divided into:

- The luminescent method;
- The color method;

- The luminescent color method;
- The brightness method; and
- The combined-penetrant NDT method (electrostatic, electric-induction, magnetic-particle, liquid penetrant-irradiation method of emission).

These methods are used to detect surface defects of metals, plastics, ceramics, and composite materials of machines, mechanisms, and devices.

21.1.2.10
Leakage NDT Method

Leakage method is widely used today to test high-pressure hydraulic mechanisms and control systems for tightness of hydrocylinders, hydromotors, valves, and other hydro-apparatuses in machines' technological systems.

21.1.3
Expert Systems

At present the most promising application of the expert system is for NDT and the technical diagnostics.

The basis for this conclusion is the analysis of their potential, the tasks, and applications. Above 50% of the world's expert systems are diagnostic. The typical problems are:

- data are not complete and they may be interpreted as having different meanings;
- data and knowledge are changing dynamically;
- large number of possible solutions;
- optimization of processing and analyzing the information necessary for making decisions, and developing the hardware and software;
- difficulties in construction of objects' models and diagnostic models;
- the necessity to choose the experienced and skillful experts;
- application of the expert in the adaptive systems of control;
- training of specialists.

If it is necessary, an expert can make the final decision.

When a technological system of machines does not have an adaptive control system, an expert carries out adaptive functions.

21.2 Qualimetry

Complex automation of the production processes called for the serial production of robots and manipulators of different types. Objective comparison between them, determination of the quality criteria and indices characterizing the conditions of use, determination of the characteristics in the certificates, choice and standardization of the parameters to be registered, and development of systematic methods for attaining and using experimental data, including those necessary for the troubleshooting of robots and their control system, are very current issues.

21.2.1 Qualimetry of Mechanisms

Qualimetry of mechanisms is based on the common theory of product-quality estimation and forms a part of the modern theory of mechanisms. One of the main objectives of qualimetry is to define the range of indices and quality criteria of mechanisms, based on their purpose and service conditions.

Qualimetry is especially significant for the automatic equipment and robotized systems, and positioning and locking mechanisms, due to the special requirements of their accuracy, reliability, and speed level. The composition of mechanisms, their structural and design characteristics, as well as dimensional and clearance characteristics of the drive and control system are also important. Besides, economic and subsidiary factors should be taken into account [7–10].

21.2.2 Quality Indices and Parameters to be Registered

The most important quality indices of the robots are the following: work envelope, velocity, rigidity, payload, accuracy of positioning and repeatability, gripping force, dynamic loads of the components of mechanisms and drives, power requirement, efficiency, amplitude-phase frequency and vibration characteristics, noise level, forces of damping, thermal fields and deformations, general power consumption, consumption of compressed air, cooling and working fluids, service life, and other characteristics of reliability. In the course of research and testing, the difficulties connected to the choice of the cycle of the robot operation must be taken into consideration. Tests may be carried out:

- during the successive implementation of the elementary components of the cycle;
- during the simultaneous implementation of several motions in most frequently occurring combinations determined by a rigid program; and
- during the simultaneous implementation of several motions of characteristic combinations and under several very difficult conditions of realization occurring in the course of working according to a flexible program (second- and third-generation robots).

Determination of the characteristic combinations of different motions to be carried out by the mechanism of the robot, especially when details or places are sought for the implementation of technological operations by second- and third-generation robots, necessitates prolonged preliminary statistical studies. During the examination and trials of robots, the following quantities must be measured and registered: time intervals (time of the cycle and its components), small and big motions, speeds, accelerations, pressures and forces, current intensity, tension, consumption of air or fluid, power, temperature, vibration parameters, and sound pressure. In a number of cases the wear of the joints and the existing contact of the mating surfaces are determined. For troubleshooting purposes, several kinematic and dynamometric parameters and signals from the control system are registered simultaneously, which makes it possible to obtain the dynamic cyclogram of the robot.

21.2.3 Standardization of Testing of the Robots' Positioning Mechanisms

The most important element of the system methods of the examination of robots is the standardization of testing. Only with the standardization of testing is it possible to decrease its labor consumption and make a useful exchange of information between research organizations, design offices, and enterprises designing, producing, and operating robots. During standardization, demands made for different types of examinations and testing must be taken into account in order to formulate the tasks of projecting and dimensioning the basic constructional units; research and development tests of the prototypes; certification tests of production models; acceptance, delivery, and check-up tests; and diagnosing of robots handed over to the customer, periodic checking, and troubleshooting tests during operation. The standardized methods of testing must make it possible to perform testing under industrial conditions. In a laboratory, the total volume of research is divided into standardized testing and additional research. When defining the places and the volume of testing, the demands of all potential users must be taken into consideration. This is a characteristic feature of the systems' approach.

Since the majority of robots used in industry include a great number of positioning mechanisms, the questions of the standardization of testing will be examined with respect to the example of the mechanisms of linear and angular positioning of the robots.

The first stage of the standardization of testing is the choice of the minimum number of initial registered parameters and relationships accessible for measuring under workshop conditions. This objective is decided by the aid of the table of the evaluation levels (Table 21.1). All the quality indices and dimension less ratings arise from the initial, experimental, calculated, or given data included in the certificate.

Depending on the number of parameters to be united and the succession of obtainment, the individual and complex indices and ratings are divided into five levels.

Indices of each successive level arise from the indices of the preceding levels or from the initial registered relations. The complex indices and ratings combine a maximum of six parameters. This enables a comparison to be made between the mechanisms of robots of different designs and sizes. The methodology of the composition of these tables is considered in more detail in the works [7–9] listed in the reference literature. The structure of the complex indices is obtained in a number of cases based on special, experimental, and industrial research.

Table 21.1 Evaluation levels on indices.

	Quality indices	
Initial	Individual	Complex
relations	Quantity of parameters to be initiated	
to be	1	2–4 2–4 3–5 5–6
registered	Levels	
	1	2 3 4 5
	Dimensionless rating	
	Individual	Complex

21.2.3.1
Mechanisms of AngularPositioning

The permitted quality of united parameters depends on the accuracy of obtaining experimental data (usually limited to four levels). An analysis and arrangement of indices of the tables composed for the mechanisms of positioning with different types of drives have shown that three parameters of the end-effector (last gleed) of the robot must be recorded for obtaining the majority of the most important indices and ratings. For example, for the mechanism of the angular positioning, such parameters are (Fig. 21.2): speed $\omega(t)$, acceleration $\varepsilon(t)$, and small displacements at the end of the movement $\delta_\psi(t)$. In addition, the force parameters such as forces, moments, pressures, etc., the selection of which depends on the type of the drive and on the design of the robot, are registered. Rotation angles and the moment of inertia of the end-effector are taken based on data given in the certificateor supplied by special measurements. The most important indices characterizing the time intervals, operation and transport velocity dynamic loads, and, indirectly, durability and reliability obtained by the recording of the three parameters are given in Table 21.2 for the mechanisms of angular positioning in the quality of the example.

The exponents in the formulae in levels 3 and 4 (n,p,k,z) are determined experimentally and need refining. For many robots it can be approximately accepted that $n = z = 0.33$; $p = 0.5$; $k = 0.25$. The notations of the time intervals and kinematic parameters are given in Fig. 21.2. The use of the velocity coefficient (K) makes it possible to compare robots with different

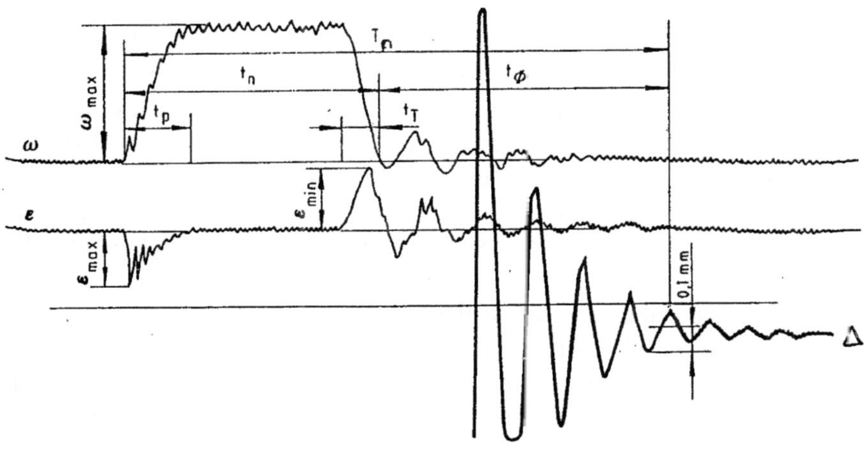

Fig. 21.2 Oscillograms of speed ω, acceleration ε, and small displacements Δ of the motion of an arm (robot with hydraulic drive).

Table 21.2 Fundamental indices of the mechanisms of angular positioning.

	Levels				
Indices	1	2	3	4	5
Time intervals	t_n, T_n t_ϕ, t_p	$t_\phi = T_n - t_n$	$\eta_\phi = t_\phi / T_n$		
Velocity	ω_{max}	$\omega_{mean_0} = \frac{\psi}{t_n}$ $\omega_{mean} = \frac{\psi}{T_n}$	$K = (\omega_{mean}) \quad J^n$ $K_0 = (\omega_{mean_0}) \cdot J^n$	$a_\omega = \frac{K}{\psi^p \delta_\psi^k}$	
Dynamic loads, durability	$\mid \varepsilon_{max} \mid$	Minmax $= J \mid \varepsilon_{max} \mid$	$K_D = \frac{\lvert \varepsilon_{max} \rvert}{\omega^2_{mean_0}}$	$K_{DB} = K_D \delta_\psi$ $A_D = K_D \delta_\psi K^z_0$	$A_{KD} = \frac{a^n_\omega}{A_D}$ $A_D = K_{DB} K_0^z$

moments of inertia J (kgm^2) of their end-effectors. Coefficients a_ω make it possible to enlarge the circle of comparable mechanisms, thus making the evaluation independent of the accepted accuracy of positioning δ_ψ (angle sec) and of the angles of rotation ψ (rad). The moment of inertia (Minmax) and the coefficient of dynamism (K_D) are calculated according to the maximum of angular acceleration ε_{max}(s^{-2}). Coefficient A_D takes into consideration, besides the coefficient of dynamism, the positioning error and velocity of the mechanism, both of which influence its reliability and durability. According to the value of the exponent z cited above, it can be seen that the greatest influence on index A_D is exerted by the coefficient of dynamism and by the accuracy of positioning. The coefficient of velocity (K_0) has a smaller effect. It is attributable to the fact that the impact of the speed of rotation is reflected by quantities K_D and δ_ψ and is taken indirectly into consideration. Depending on the level of vibration at the end of the rotation, the accuracy of positioning is interconnected with the level of dynamic loads in the swivel mechanism and, if stops are present, also with the mechanism of fixation. The case is similar to the reliability of the operation of the positioning mechanism, which essentially affects the general reliability of the operation of the positioning mechanism, which essentially affects the general reliability of the robot.

It appears from the analysis of these complicated relationships that for the standardization of this coefficient and for the refinement of the structure of the relationship of A_D with the different parameters lengthy investigations have to be made in the process of operation. However, analysis of experimental and calculation data shows that the designer's attention must be called to quantity A_D and he must be warned to be cautious in making decisions. The value of A_D may signal the incorrect adjustment of the

Table 21.3 Fundamental indices of the mechanisms on linear positioning.

Indices	Levels				
	1	2	3	4	5
Time intervals	t_n, T_n t_p, t_T	$t_\Phi = T_n - t_n$	$\eta_\Phi = t_\Phi / T_n$		
Velocity	v_{max}	$v_{mean_0} = \frac{L}{t_n}$ $v_{mean} = \frac{L}{T_n}$	$K_{l_0} = v_{mean_0} m^n$ $K_l = v_{mean} m^n$	$a_v = \frac{K_l}{L^p \Delta_l^K}$	
Dynamic loads, durability	$\lvert a_{max} \rvert$	$P_{in} = m \lvert a_{max} \rvert$	$K_{Dl} = \frac{\lvert a_{max} \rvert}{v^2_{mean_0}}$	$K_{DlB} = K_{Dl} L$ $A_{Dl} =$ $K_{Dl} L \cdot K_{l_0}^Z \Delta_l$	$A_{KDl} = \frac{a_v^m}{A_{Dl}}$ $A_{Dl} =$ $K_{DlB} \cdot K_{l_0}^Z \Delta_l$

mechanism or inadmissible wear of the components of the manipulator or of the robot. The method of measuring first-level parameters is illustrated in Fig. 21.2.

Time t_n is determined (in seconds) by the aid of the first point of intersection of the oscillogram of angular speed with the axis of the abscissa. Time T_n is determined by recording $\underline{\omega}$ and $\underline{\delta_\Psi}$ with the given quantity ($\underline{\delta_\Psi}$ in angular seconds). In Fig. 21.2, this corresponds to the vibration allowance of the robot's hand equaling 0.1 mm. It appears from Fig. 21.2 that in the case of such strict limitation of the value of vibration, $T_n > 2t_n$. The time losses on the damping of vibration (t_Φ) are taken into consideration with the aid of coefficient η_Φ (Table 21.2), essentially depending on the acepted δ_Ψ. In such a way the connection of T_n and η_Φ with δ_Ψ is taken into account along with the methods of performing the experiment, which represents its characteristic feature. Investigating robots with hydraulic or pneumatic drive, the pressure in both chambers of the motors is also recorded.

21.2.3.2
Mechanisms of Linear Positioning

The most important indices characterizing the mechanisms of linear positioning are given in Table 21.3.

In Table 21.3 v_{max} [m/s], a_{max} [m/s^2] – the maximum of linear velocity and acceleration, P_{in} – force [kgm/s^2]; K_{Dl} – coefficient of dynamism [1/m]; $K_{Dl} \cdot L = K_{DlB}$, K_{l0}; K_l ; a_v -velocity coefficients, m [kg], Δ_l – accuracy of linear positioning [mm]. The exponents in the formula are determined experimentally: $n = m = 0.33$; $p = 0.5$; $k = 0.25$; and need refining.

21.2.4
Analysis of the Most Important Characteristics of the Intermittent Motion Mechanisms

Evaluation of the primary and complex indices is used for checking the conformity of the fundamental characteristics of the robot with the data of the certificate. In this connection, the certification values of accuracy of δ_ψ, and rotation time T_n, the maximum value of acceleration ($a\varepsilon_{max}$), etc., are accepted as quality norms. The question is more complicated if speeds are compared. The certified quantities do not consider the dependence of the mean speed value on the length or angle of displacement, which, in turn, does not make it possible during the elaboration of adjustment to calculate the time of cycle T_c accurately. It is expedient to complete the certified data with the experimentally determined dependence of speed on the angle of rotation.

This dependence can be used to good advantage during the evaluation of operation velocity. When comparing robots differing from one another as regards to the value of J and δ_ψ, such charts are not sufficient, and so it is necessary to standardize values K and a_ω which are used for the evaluation of velocity. Although this evaluation is only an approximate one, it can make it possible to avoid great errors in the course of design. The charts in Fig. 21.3 show that for all the investigated robots, it is indispensable to take function $\omega_{mean} = f(\psi)$ into account.

Robots with electromechanical drive (curves 1, 2, 3 in Fig. 21.3) are distinguished by their low speed ($\omega_{mean} < 0.4–0.8\ s^{-1}$).

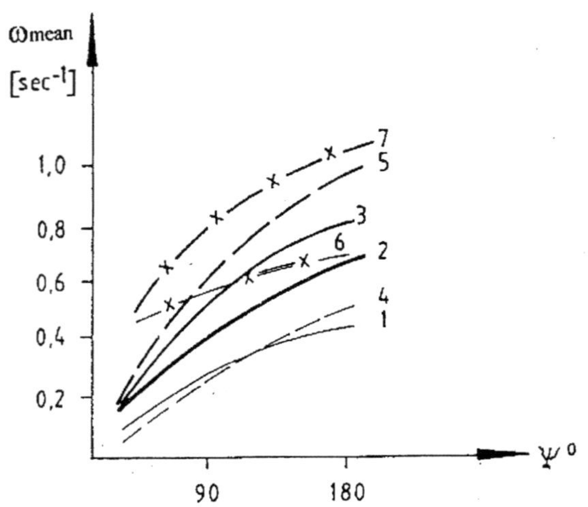

Fig. 21.3 Dependence of mean speed ω_{mean} on the trajectory (angle φ) for robots.

Robots with hydraulic drive (curves 4 and 5 in Fig. 21.3) differ considerably in respect of speed. It is due to the difference between the mass of parts to be moved (10–130 kg). In case the parts to be transferred are of a smaller mass, speed can attain $\omega_{\text{mean}} = 1\ \text{s}^{-1}$.

Robots with pneumatic drive (curves 6 and 7 in Fig. 21.3) for parts weighted 5 to 10 kg excel as a result of their high speed (in a number of cases $\omega_{\text{mean}} > 1\ \text{s}^{-1}$). The character of the function $\omega_{\text{mean}} = f(\psi)$ for the robot demonstrated by curve 6 in Fig. 21.3 testifies to its malfunction.

The most important characteristics of the intermittent angular motion mechanisms are:

$$\omega_{\text{mean}} = f(J);\ K = f(\psi);\ K_D = f(K0);\ \eta_\Phi = f(\psi);$$
$$a_\omega = f(A_D);\ A_{KD} = f(A_D);\ K_B = a_{\omega B} = \sqrt[3]{10}$$

The most important characteristics of the intermittent linear motion mechanisms are: $v_{\text{mean}} = f(L)$; $v_{\text{mean}_0} = f(m)$; $K/K_B = f(\Delta_l)$; $K_{Dl} = f(K_{l0})$; $a_v = f(A_{Dl})$ [7].

21.3
Qualimetry Data Application for Diagnostics

For diagnostics we use:

- border values of complex indices (see Fig. 21.4);
- the tables of experimental data (Tables 21.4–21.6);
- the dependence of parameters and indices on wear.

Experimental research of automatic equipment in operation showed the decrease of the complex index a_ω and increase of A_D, characterizing durability. The permitted change of parameters is limited by $A_{KD}/A_{KDB} < 10^{-6}$.

By average $A_D/A_{DB} = 2000$; $A_{KD}/A_{KDB} = 4 \cdot 10^{-4}$.

The dependence of K_D on indices K_0/K_B and δ_ψ for different mechanisms was studied experimentally (Table 21.4). Speed level (index a_ω) depends on unit type, indexing, and locking mechanisms (Table 21.5).

Average data for K/K_B by $\psi = \pi/3$ and high, average, and low positioning accuracy are given in Table 21.6. For another ψ the data may be calculated for application in diagnostics.

The dependence of η_Φ on angle ψ is found experimentally. By small ψ indices, η_Φ is high. High t_Φ and η_Φ are by indexing mechanism with gear-coupling ($\delta_\psi < 5''$). Low t_Φ and η_Φ are on the mechanisms with a single fixator ($\delta_\psi > 20''$).

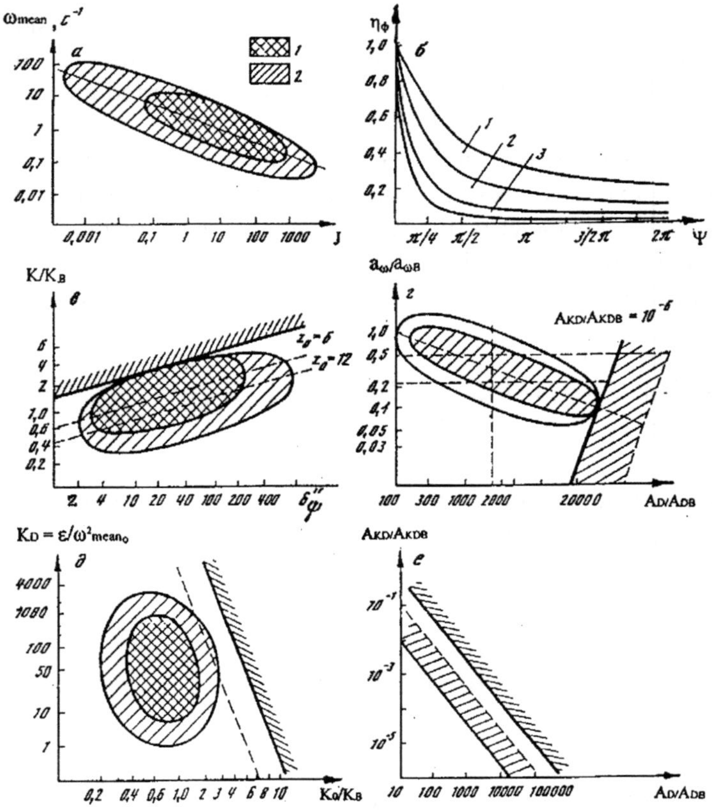

Fig. 21.4 The main dependence between fundamental indices and parameters: 1 – region of the most frequently supervised parameters; 2 – region of extra data.

Table 21.4 Coefficients K_0 and K_D characteristic on different types of mechanisms.

Type of mechanisms	K_0 / K_B	K_D	δ''_{ψ}
Electromechanic cams	0.5–3	2–55	50–400
Geneva mechanisms	0.25–3.5	4–200	15–300
Lever-and-ratchet	1–3.5	20–90	40–200
Gear-and-ratchet	0.2–6.0	1–20	200–1000
Gear, wave	0.3–6.0	5–400	5–170
Pneumomechanic	0.35–3.2	0.8–60	100–900
Hydromechanic	0.32–3.0	150–1000	4–20
Total for this sample	0.2–6.0	0.8–1000	4–1000

Table 21.5 The dependens between indices η_Φ and the angle ψ.

Type of mechanisms	ψ, rad				
	$\pi / 16$	$\pi / 8$	$\pi / 4$	$\pi / 2$	π
	η_Φ				
Singl fixcator (type I)	0.4–0.6	0.2–0.45	0.1–0.25	0.03–0.1	0.01–0.05
With revers (type II)	0.6–0.7	0.45–0.65	0.25–0.45	0.1–0.28	0.05–0.15
Gear-cupling (type IV)	0.7–0.95	0.65–0.9	0.45–0.7	0.28–0.45	0.15–0.35

Table 21.6 Speed level coefficients K / K_B by $\psi = \pi/3$.

Positioning accuracy	δ''_ψ	1	2	3	4	5
High	K / K_B	0.5	0.56	0.66	0.7	0.75
	$\sqrt[4]{\delta_\psi}$	1	1.19	1.32	1.41	1.5
	δ''_ψ	6	7	8	9	10
	K / K_B	0.78	0.81	0.84	0.86	0.89
Average	$\sqrt[4]{\delta_\psi}$	1.56	1.63	1.68	1.73	1.78
	δ''_ψ	20	30	40	50	60
	K / K_B	1.05	1.17	1.26	1.33	1.4
	$\sqrt[4]{\delta_\psi}$	2.1	2.34	2.53	2.66	2.8
	δ'_ψ	1.5	2	3	4	5
Low	K / K_B	1.54	1.65	1.83	1.92	2.08
	$\sqrt[4]{\delta_\psi}$	3.08	3.3	3.67	3.67	4.15

21.3.1 Qualimetry Data on Locking Mechanisms

Complex automation of the production processes and complicated automatic equipment of technological systems call for the locking mechanisms of different types [7].

In automatic equipment, locking mechanisms with a single fixator (I,II,III,IV type) or with two fixators (II type) are used. The best are fixators without slide contact. Fixators with slide contact during indexing or working of the automatic equipment are not recommended for a new precision construction.

Single fixator with slide contact by locking (I,IV) in many devices has compensation-of-wear construction. A shape of surfaces is of great importance (cylindrical, taper, flat, screw, and so on). Many locking mechanisms with compensation of wear have clutch (closing) mechanisms.

In flexible manufacturing systems, many locking mechanisms with gear coupling ($z \leq 360$) are used. Compensation of errors and wear provides high positioning accuracy ($\delta_\psi = 1$–$10''$). These mechanisms (IV type) are provided with preliminary locking. The Geneva mechanisms are also provided with preliminary locking.

Indexing mechanisms with clutch mechanisms (IV type) and locking mechanisms with reverse (Table 21.5, II type) have gross time distributions – t_Φ and η_Φ.

21.3.2
Qualimetry Data on Index Tables

Qualimetry data on index tables are investigated by:

- experimental research (laboratory works, research in industry); or
- computer-aided research (mathematical modelling).

Computer-aided research cuts the design time and sets higher standards in advanced product design. In computer-aided research the data of experimental research are used.

The mathematical model of the indexing mechanism with hydraulic drive (Fig. 21.5) is shown in Fig. 21.6. The indexing mechanism works with reverse on the table (locking mechanism – type II); ψ_{ox} – angle of reverse.

In simulation, the dynamic characteristics of this structure were described by the following equations [7]:

$$\ddot{\psi}_1 + \frac{b_1}{J_1}(\dot{\psi}_1 - \dot{\psi}_2) + \frac{C_1}{J_1}(\psi_1 - \psi_2) - \frac{C_g}{J1}\int_0^t (\dot{\psi}_H - \dot{\psi}_1)d\tau = 0$$

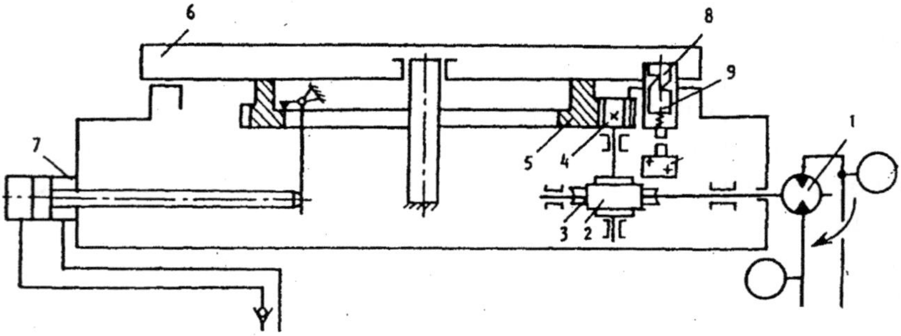

Fig. 21.5 Indexing table with hydraulic drive: 1 – hydromotor; 2 – worm; 3 – worm gear; 4 and 5 – gears; 6 – table; 7 – hydrocylinder; 8 and 9 – locking pins.

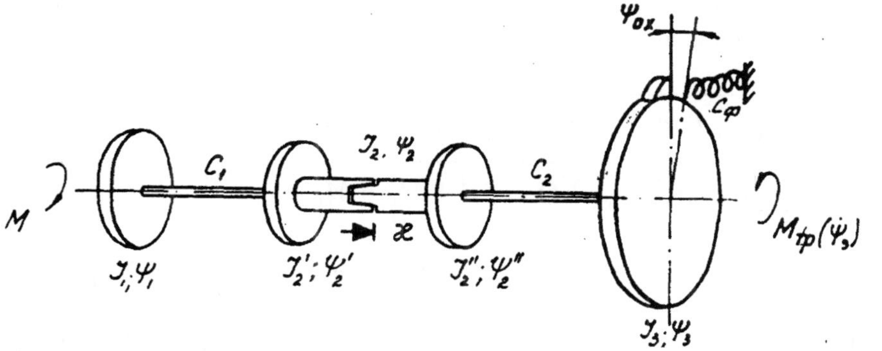

Fig. 21.6 Mathematical model of the indexing mechanism with hydraulic drive.

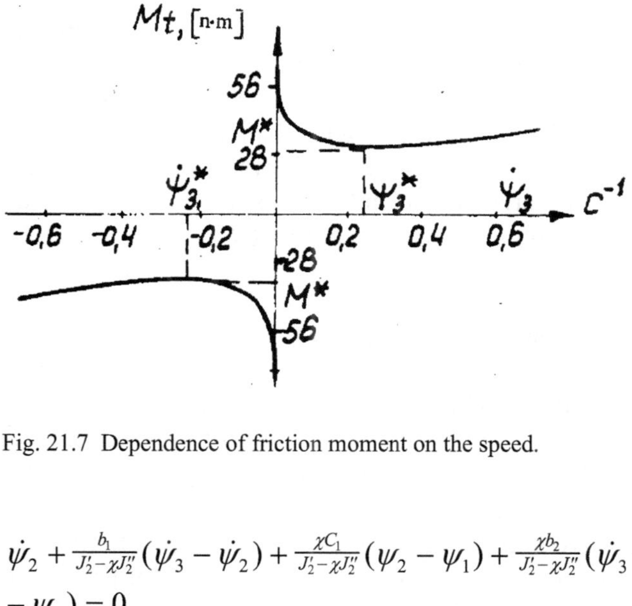

Fig. 21.7 Dependence of friction moment on the speed.

$$\dot{\psi}_2 + \frac{b_1}{J_2' - \chi J_2''}(\dot{\psi}_3 - \dot{\psi}_2) + \frac{\chi C_1}{J_2' - \chi J_2''}(\psi_2 - \psi_1) + \frac{\chi b_2}{J_2' - \chi J_2''}(\dot{\psi}_3 - \psi_2) + \frac{\chi C_2}{J_2' - \chi J_2''}(\psi_3 - \psi_2) = 0$$

$$\ddot{\psi}_3 + \frac{b_2}{J_3}(\dot{\psi}_3 - \dot{\psi}_2) + \frac{C_2}{J_3}(\psi_3 - \psi_2) + \frac{C_\phi}{J_3}(\psi_3 - \psi_{ox}) + \frac{M_1}{J_3} = 0$$

where: J_1, J_2, J_3 = reduced moments of mass inertia; C_1, C_1, C_Φ, C_g= coefficients of rigidity; ψ_1, ψ_2, ψ_3, ψ_{ox} = angles; $\dot{\psi}_1, \dot{\psi}_2, \dot{\psi}_3, \dot{\psi}_H$ = speeds; $\ddot{\psi}_1, \ddot{\psi}_2, \ddot{\psi}_3$ = accelerations; $M(t) = f(\psi_3)$ = braking torque (Fig. 21.7); M = reduced moment of hydraulic drive (Fig. 21.6).

The definition of dynamic transmission ratio χ is given in [7]. Locking forces are calculated by:

$$Q_{\max} = \tfrac{1}{R_\Phi}\sqrt{C_\Phi J_3(\dot{\psi}_{\text{ox mean}})^2(1+\tfrac{\delta_\omega}{2})^2 + M_t^2} - M_t$$

$$Q_{\min} = \tfrac{1}{R_\Phi}\sqrt{C_\Phi J_3(\dot{\psi}_{\text{ox mean}})^2(1-\tfrac{\delta_\omega}{2})^2 + M_t^2} - M_t$$

and $\delta_\psi = \Delta Q \cdot R_\Phi / C_\Phi = (Q_{\max} - Q_{\min}) \cdot R_\Phi C_\Phi$ where R = radius of the pin displacement [7].

In calculation we accepted: $\delta_\omega < 0.3$; $Q_{\min} > 2000$; $Q_{\max} < 41000$ n (by condition of durability and reliability).

The quality indices calculated by simulation results for $z = 4$ are: by $\delta_\psi'' = 5 - 26$; $T_n = (2.4–4.7)c$; $K_0 / K_B = 1–2.2$; $K / K_B = 0–94–2$; $a_\omega / a_{\omega B} = 0.42–0.9$.

Experimentally the tables with $z = 4–6$, $D = 0.8–1.25$ m were examined. For $z = 4$: $J = 30–540$ kg m^2, angular acceleration $\varepsilon_{\max} = (2.7–12)c^2$; $K_D = 150–500$; $K_0/K_B = 0.56–1.2$; $a_\omega / a_{\omega B} = 0.23–0.53$; $A_D / A_{DB} = 2400–12300$. The experimentally received time T_n was higher than during simulation, because of design failures (low ω_{mean}). Fig. 21.8 shows some oscillograms obtained in experimental research of the indexing table (Fig. 21.5) with hydromechanic drives and clamping devices = pressure P_3.

21.3.3
Qualimetry Indices and Parameters on a Turret

For many machine tools, the machining time is smaller than transport, positioning, and waiting time. The positioning time of a turret is likely to be reduced. The standards for metal-working tools have been raised to great speeds of technological development.

CNC machine tools have higher precision and speed by positioning of a turret. Most of the turrets have a gear-coupling mechanism for precision locking and must have a preliminary locking mechanism that orientates the teeth of the gears before coupling. At great speeds, the preliminary locking mechanism is a weak link of the indexing mechanisms.

Reliability analysis was made by laboratory experiments and by exploitation (turret with electromechanical and hydraulic drives).

Qualimetry indices of a big turret on a CNC machine tool with hydraulic drive are: $a_\omega / a_{\omega B} = 0.4–0.6$; $A_D / A_{DB} = 85–295$; $(A_{KD}/A_{KDB}) \cdot 10^3 = 2.4–10$.

Preliminary locking is carried out by two rollers of Geneva mechanisms' disks.

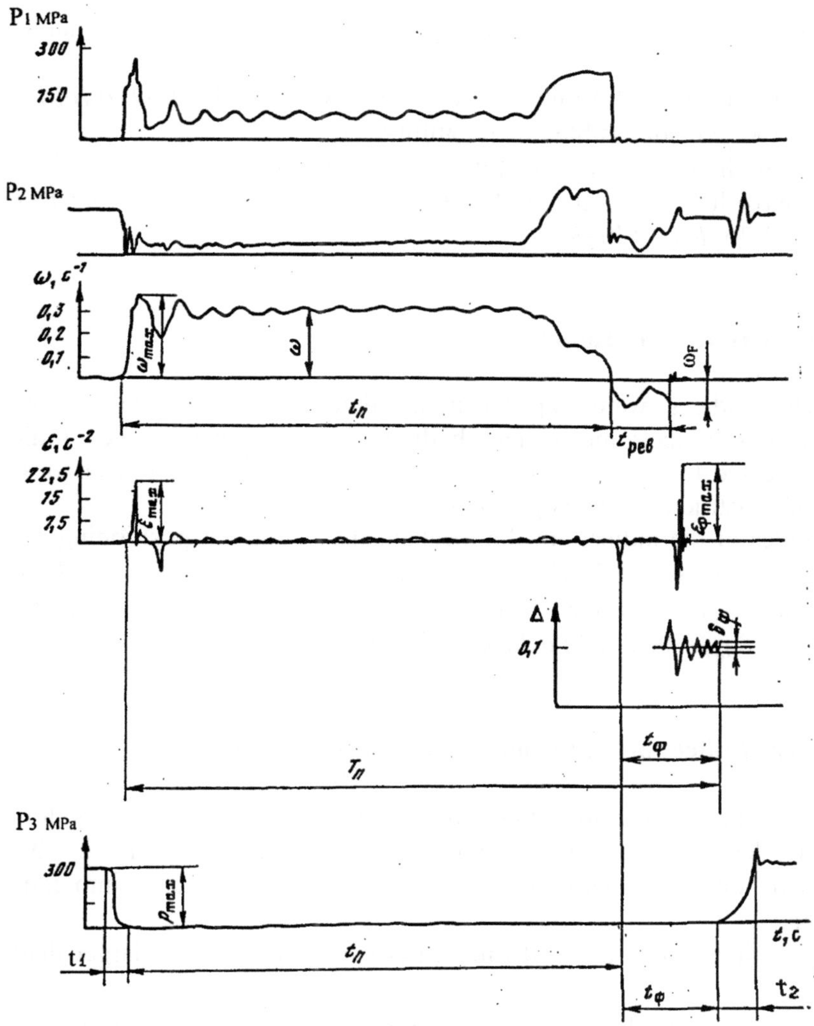

Fig. 21.8 Kinematic and pressures' oscillograms for indexing table: P_1, P_2, P_3 – pressures in hydrosystem; ω – speed; a – acceleration, Δ – small motion, δψ – accuracy of positioning; t_1 and t_2 – clamping and setting-free time intervals, ω_F – speed of locking.

Qualimetry indices on turret-preliminary locking mechanisms are: $a_{\omega_0}/a_{\omega B} = 0.1–0.12$; $(A_{D0}/A_{DB})\cdot 10^{-4} = 2.2–6.1$; $(A_{KD0}/A_{KDB})\cdot 10^5 = 0.82–6.7$ where: δ_p = accuracy of angular positioning at preliminary locking; a_{ω_0} = speed coefficients at preliminary locking; $A_{DB} = K_D \delta_p \sqrt[3]{K_0}$; $A_{KD0} = a_{\omega_0} / A_{D0}$.

By comparing data: $a_{\omega_0} < a_\omega$; $A_{D0} >> A_D$ and $A_{KD0} >> A_{KD}$.

This confirms the fact of preliminary locking mechanisms' low reliability.

A turret with n_0 =210 r/min; δ_p = (22–65)″; K/K_B = 1.44–1.59; $a_\omega/a_{\omega B}$ = 1.35–1.06 has non admissible low reliability.

Turrets with electromechanical drive (planetary, worm, and gear reducers) also have low reliability by J = (0.35–1.2) kg m^2; t_n = 0.6–1–3c; K_0 / K_B = 0.67–1.76; K_D = 140–320.

21.3.4
Qualimetry Indices on Robots

Qualimetry indices were experimentally determined for robots (first-generation) with electromechanical, hydraulic and pneumatic drives. Indices are directed to a more precise definition of certificate characteristics, defects of construction, and adjustment of robots.

Examples of the indices for the mechanisms of angular positioning robots are: K_0 / K_B = 0.1–1.7; K/K_B= 0.08–1.2; $a_\omega/a_{\omega B}$ = 0.04–0.56; (A_D / A_{DB})· 10^2 = (2.6–864) , and for mechanisms of linear positioning – (m=5–100 kg): $K_{l_0}/K_{l_0 B}$ = 0.13–2.8; a_V/a_{VB} = 0.5–10; A_{Dl}/A_{DlB} = 0.1–18.

21.3.5
Qualimetry Indices on Transporters of Transfer Lines

Transporters move sometimes big masses of large parts of motor-cars, lorries, and tractors. With large weight m and linear movement L, the dynamic loads are also large: Q=(28–65)·10^3 [n] and must be taken into consideration.

Acceleration is low: a = 3–10 m/c^2, but indices K_{l_0}, A_{Dl} are larger than for robots.

Experimental speeds and acceleration for one of the transporters are given in Fig. 21.9. Many transporters have hydraulic drive and high speed v, $K_{l_0}/K_{l_0 B}$ = 1.8–10; a_V/a_{VB} = 1.1–9.5; A_{Dl}/A_{DlB} = 42–410.

21.4
Complex Methods of Diagnostics With Standard Oscillograms Measuring Cyclic Automatic Equipment

Complex methods of diagnostics include:

- packaged approach;
- use of mathematical simulation, computer modeling;

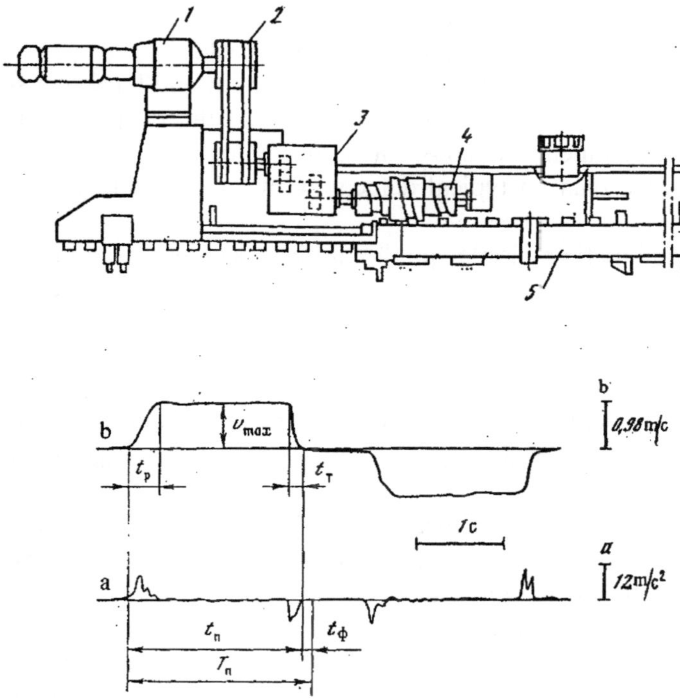

Fig. 21.9 Transporter with cam drive: 1– electromotor, 2 – belt drive, 3 – reducer, 4 – cam, 5 – transporter, a – acceleration, b – speed.

- use of standard oscillograms, defect charts, monitoring;
- use of qualimetric methods;
- use of organoleptic methods;
- diagnostics of automatic equipment, technological process, tools, workpieces, technological surroundings, waste materials, and environment.

21.4.1 Packaged Approach, Procedure, and Mode of Experiments

Packaged approach includes:

- standardization of testing and monitoring automatic equipment at the different stages of the life cycle;
- co-operation of experts, and exchange of information between them;
- creation of diagnostic systems, software, transducers, and database;
- exchange of knowledge and information in the education, training, and publishing of conferences, seminars, discussions, and exhibitions.

Procedure and mode of equipment are made on examples of automatic multiple-spindle-turning machines and flexible turning cells.

Several hundreds of turning machines in the motor-car and lorry industry, in bearings' manufacturing were investigated, and the database was created for use with qualimetry methods, standard oscillograms, and charts.

For many years the data on typical defects, wear, and reliability have been collected.

On the stands, automatic multiple-spindle-turning machines were investigated by various speeds, regulation of mechanisms, and control systems, by machining various parts. Mathematical simulation of a mechanism's dynamics was made.

Assembly defects were investigated at manufacturing; standard oscillograms and charts were obtained. Some of the tasks are monitoring, testing, diagnostics of every automatic type of equipment, and personnel training.

By operation of automatic equipment we have investigated reliability, type of defects, and influence of wear on dynamic characteristics; reviewed the parameters that will be determined, and have developed standard charts.

Under repair, the wear of parts was measured and the equipment after assembly or modernization was diagnosed. By life investigation and certification, qualimetry methods and simulation methods are used for more precise forecasting of reliability and investigated certification data.

21.4.2 Standard Oscillograms and Defect Charts

Standard oscillograms and defect charts make easier diagnostics for personnel in manufacturing and operating conditions.

They were prepared beforehand in the laboratory on stands and were updated by manufacturing and operations. In diagnostics oscillograms of velocity, acceleration, pressures (Fig. 21.8), forces, moments of torsion, and current intensity are used.

For automatic multiple-spindle-turning machines, moment of torsion is used, measured on the camshaft (Fig. 21.10). On these oscillograms, three time periods are portioned: I– locking pins' take-out; II – turn of multiple-spindle trommel; III – locking. On oscillograms several defects are detected: (1) gross moments during locking pins' take out; (2) incorrect regulation of locking mechanisms' take-out; (3) incorrect assembly of trommel turn mechanisms; (4) insufficient time of working of turn mechanism on assembly stand; (5)incorrect assembly of trommel-lifting mechanisms (before turning); (6) incorrect assembly of the support's locking mechanism.

In defect charts the advice for errors elimination is given.

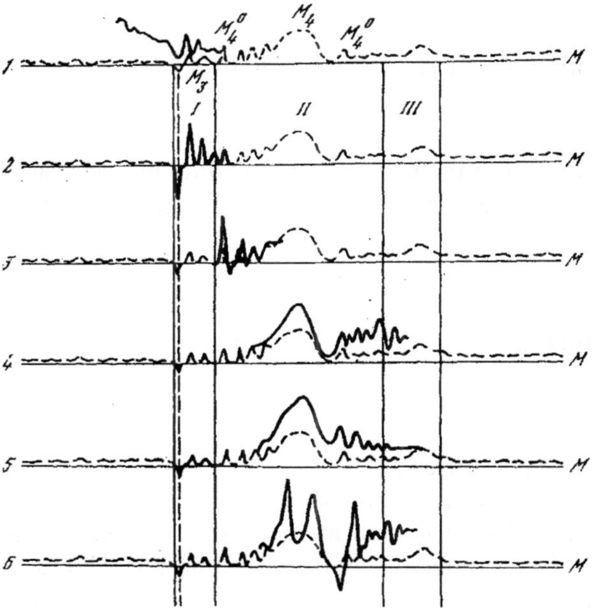

Fig. 21.10 Standard oscillograms of the moments measured on camshaft.

21.4.3
Estimation of a Mechanism's Quality and Condition by Qualimetry

The quality and condition of mechanisms are estimated:

- by moments of torsion;
- by velocity, acceleration, and small motion;
- by complex indices.

For mechanisms of multiple-spindle-turning automation, simple and complex indices are: $\omega_{\text{mean}} = (0.45–1.3)c^{-1}\varepsilon_{\max} = (1.2–120)c^{-2}$; $K_D = 16–165$; $a_\omega / a_{\omega B} = 0.3–0.86$. Indices $A_{KD}/A_{KDB} = (0.56–3.3)\cdot 10^{-3}$ are found in a favorable zone (Fig. 21.4). This is confirmed by the data of reliability indices.

21.5 Investigation of Mechanisms' Wear at the Stage of Their Manufacture and Operation

The wear of mechanisms and its influence on the dynamic and accuracy parameters is investigated during assembly, application, and repair (Scherbakov 1998).

21.5.1 At the Stage of Manufacturing

At this stage we investigate:

- wear influence on dynamic parameters;
- time period of shaking down during assembly.

The change of the maximum value of torsion moment M_{max}, measured on an assembly stand by wear of a trommel's support, is given in Fig. 21.13a.

For 5 hours M changes from 1300 to 800 nm, M intensively decreasing for 2 hours.

The time of shaking down depends on quality of parts and the assembly process and is unique for every unit of equipment.

21.5.2 At the Stage of Operation

At this stage we investigate:

- geometrical parameters of the machine parts after wear;
- wear influence on dynamic and accuracy parameters;
- shaking down period.

For intermittent angular-motion mechanisms, the wear depends on the moment of inertia J, speed ω_0 of the drive shaft, moments of friction forces M_t, and size of backlash.

In many cases torsion moments on automated camshafts change the sign. This depends on the value of coefficient $A = J\omega_0^2 / M_t$ [7].

For Geneva, drive $A = (1-2\lambda\cos\varphi + \lambda^2)^2 / \lambda(1-\lambda^2)\sin\varphi = f(\varphi, z_k)$, where $\lambda = \sin(\pi/z_k)$; φ = angle of drive shaft, z_k = the number of position of Geneva drive.

$A = f(\varphi)$ for $z_k = 6$ are given in Fig. 21.11.

For A_{II}, moment M on the camshaft changes twice. This is not good for wear condition. In Fig. 21.12, the wear of the Geneva drive cross-plate is given. The wear Δ of the main plate (by $M > 0$) is smaller ($\Delta = 0.6–0.7$ mm) than of another plate ($\Delta = 0.84–1$ mm). This depends on the size of the backlash and coefficient A_{II}. By points φ_1 and φ_2 (Fig. 21.11) we can calculate moments of friction forces: $M_t = J\omega_0^2/A$.

We can investigate the points φ_1 and φ_2 by use of impress from plates' surfaces (after wear).

The change of the maximum value of torsion moment M_{max} on a camshaft at wear of a trommel's support during application of the automatic equipmentis given in Fig. 21.13b. Moment M_{max} changes from 800 to 370 nm over 3 years. The intensive decrease of M_{max} continues for 2 years.

During application, the change of M influences the dynamic forces and standard oscillograms, and for diagnostics we must prepare the same several standard oscillograms and tolerances as for measured parameters.

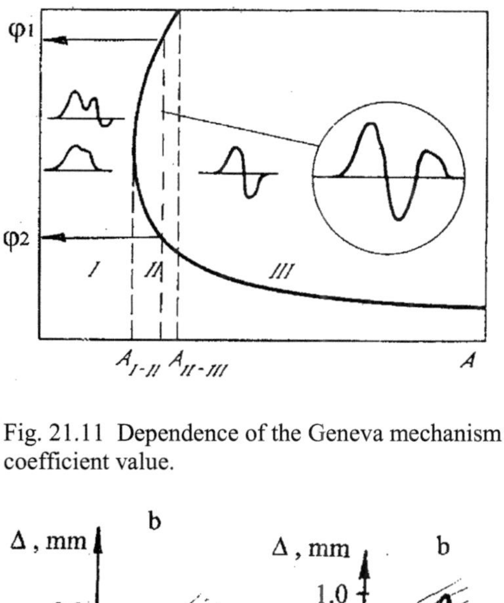

Fig. 21.11 Dependence of the Geneva mechanism drive shaft's torsion moment on the A coefficient value.

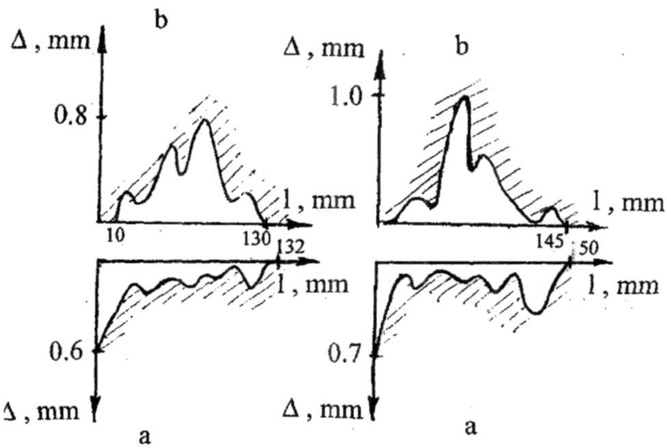

Fig. 21.12 Wear of Geneva drive cross: a – main plate; b – another plate.

21.6 Monitoring and Diagnostics of Technological Process.

21.6.1 Introduction

The main diagnostic methods and analyses of technological processes are:

- nondestructive testing of materials, workpieces, cutting tools;
- analysis of environmental influences;

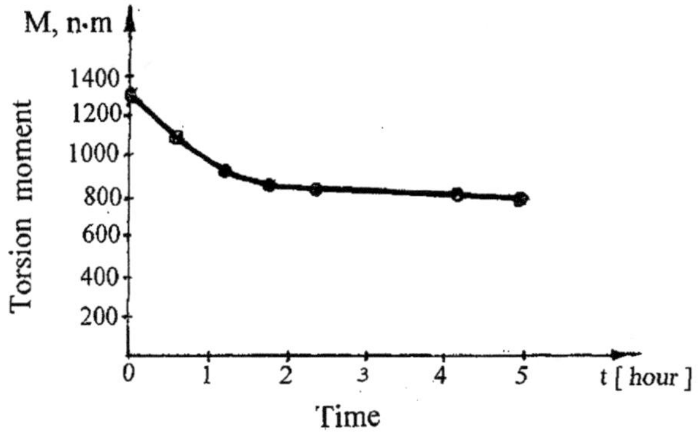

a

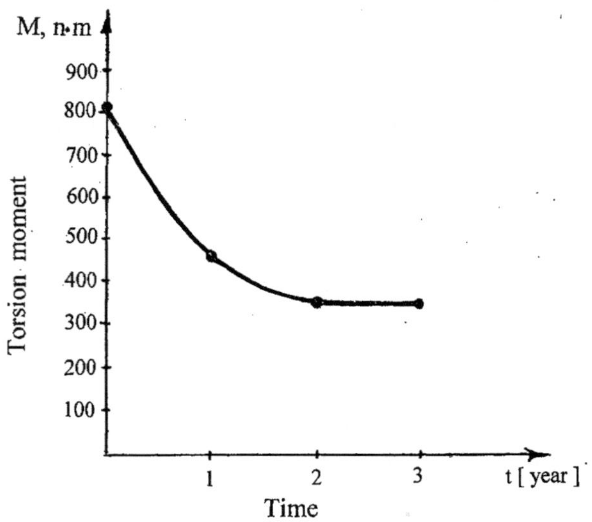

b

Fig. 21.13 Wear of trommel's support of automatic multiple-spindle-turning machines: dependence of maximum torsion moments on time: a – on assembly stand; b – during operation.

- analysis of quality and diagnostics with the help of quality indices and standard oscillograms, monitoring of dynamic forces and precision parameters;
- analysis of temperature fields and deformation;
- analysis of elasticity, deformation, vibro-acoustic diagnostics of machines and elements of technological systems;
- diagnostics of trouble-shooting;
- specific methods of diagnostics of laser, electron beam, plasma technologies, and nanotechnologies;
- use of diagnostic parameters for purpose of adoption of intelligent control systems;
- safety analysis, environmental diagnostics, and prevention of the danger of radiation and chemical emergencies.

With the help of this method, the following tasks are determined: testing and diagnostics; control, forecasting, and prevention of damaged production; increasing quality, productivity, and reliability.

21.6.2 Nondestructive Testing of Materials, Instruments and Workpieces

The NDT methods are used for determination of:

- internal condition, mechanical residual stresses, cracks, and foreign object inclusions;
- geometry defects: diameter, thickness, and other dimensions of workpieces or instruments;
- quality and destruction of surfaces: roughness, waviness, cracks, or scratches;
- structural changes in materials, physical and chemical properties, and physical-mechanical characteristics.

The NDT is based on physical processes of interrelation between a physical field and a substance with materials.

21.6.3 Diagnostics of Technological Process During Operation

The main tasks of this phase of the life cycle are:

- testing of the cutting tools' wear, forecasting troubles and errors caused by wear of a cutting tool;

- testing errors caused by thermal deformation of a technological system;
- testing errors caused by geometrical deviation of the equipment;
- testing errors caused by elasticity of a technological system;
- adaptive control of regimes;
- testing the quality of prepared products;
- testing waste materials;
- testing environmental parameters (when it is necessary).

21.6.4 Testing of Technological Inheritance

The task of testing is to ascertain how errors are formed during the whole technological process and to influence the structure of a technological process, by excluding conditions of formation of errors of different kinds.

21.6.5 Testing for Solving the Problems of Precision Engineering

Tolerances for measured parameters of prepared production are dependent on quality of measuring machines, instruments, and environmental parameters.

Systematic mistakes on a co-ordinate measuring machine (CMM) include: mistakes of measurement, deviation of position (of tables, turrets, IR, and so on), and mistakes of working out the results, analysis, and interpretation of information by an operator (the human factor). Tolerances are smaller when orientation of measured parts is automated by using perfect instruments for measuring and calibration.

Using CMMs for measuring size, form, and relative position of surfaces of complicated parts in testing (and certification) is very effective and gives perspective for creation of adaptive systems (modules, centers, and laboratories).

The productivity of CMMs is influenced by measuring speeds and by software. CMMs can make automatic measurements during the production process and provide information for the processes' correction.

21.7 Diagnostic Methods for Testing Machine Units with the Help of Built-in Devices

The most important feature of the preparation for diagnostic methods research is fault analysis of similar machine constructions during exploitation.

In the molding lines, there are most significant defects of pumps, hydrocylinders, and valves due to their slip and internal wear. In hydromechanical devices, defects also appear due to transmission mechanisms' wear.

There is a schematic of the drive of the molding machine of the indexing table illustrated in Fig. 21.14, which consists of hydrocylinder 1, gear 2, crank 3, and slide 4. Slide 4 links with the table of hydrocylinder 5.

In Fig. 21.14 ω and ε (transducers D1 and D2, Table 21.7) = velocity and acceleration of the table, P_n and P_χ = high and low pressures in the cylinder (transducers D3 and D4), P_u = pressure at the pump exit (transducer D6). As a result of that research there have been chosen the places for setting transducers that could work into construction periodically or continuously and also the places for periodic manual setting of transducers.

The next stage of diagnostic methods research is design or choice of transducers and working out the guidance for volumes and periods of diagnostics, diagrams, curves, tolerances for various parameters, etc. (Table 21.7).

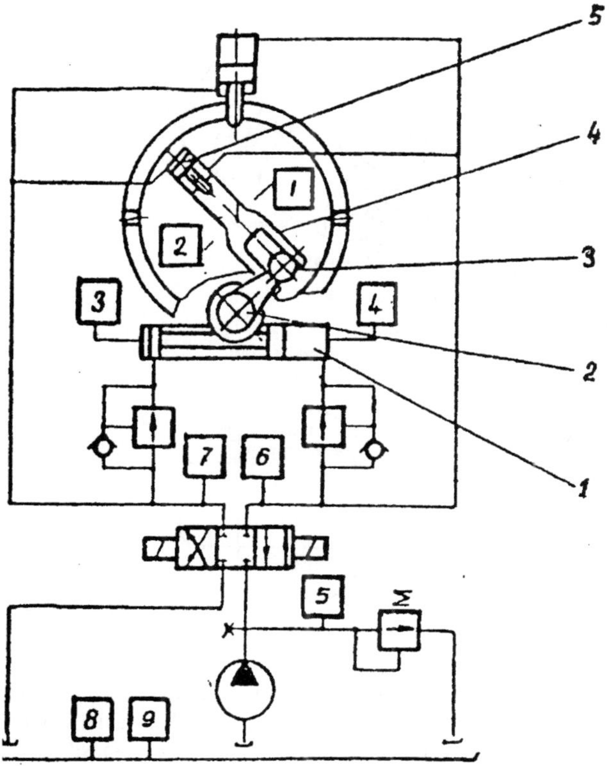

Fig. 21.14 Schematic of the indexing table's hydraulic drive (molding machine).

Table 21.7 Diagnostic parameters and testing equipment (Molding line).

Parameters for testing	Testing equipment	Construction changes	Period of testing
Pump coefficient of performance	Hydrotester	Get place for testing by D10	2 weeks
Table velocity	Velocity transducer D1	-	3 months
Table acceleration	Acceleration transducer D2	-	3 months
Pressure	Pressure transducers D3, 4, 5	transducers built in construction	3 months
Presence of pressure	transducer with indicators D6, 7	transducers built in construction	permanent
Oil level	transducer D8	transducers built in construction	permanent
Oil temperature	transducer D9	transducers built in construction	permanent
Cycle duration	time relay	transducers built in construction	permanent

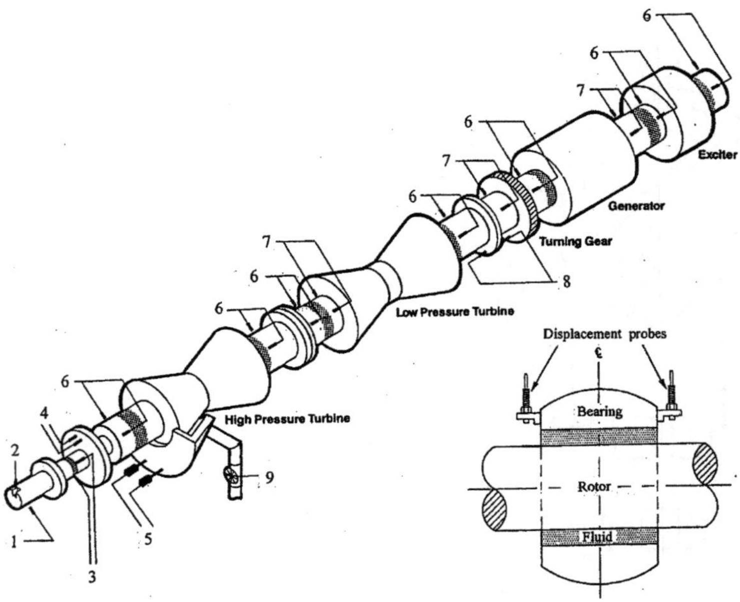

Fig. 21.15 Diagnostics of high pressure turbine: 1 – eccentricity probe; 2 – key phasor probe; 3 – speed/rotor acceleration probes; 4 – dual-voting axial-thrust probes; 5 – dual-case expansion transducers (LVDTs); 6 – dual-radial vibration XY probes; 7 – dual-radial vibration XY-mode identification probes; 8 – differential-expansion probes; 9 – valve-position transducers (potentiometer).

In rotating machinery diagnostic systems, a monitoring system with many transducers (Fig. 21.15) built into the device ([11] Bently Nevada) includes: displacement, speed, acceleration, pressure, vibration, eccentricity, key-phasor, valve position, torque moments transducers, and others.

21.8 Evaluation Methods of NDT and Diagnostics and the Perspectives of Their Application in Industry

The NDT and diagnostic methods are applied at all the phases of design, manufacturing, operation, and audit of the plant.

21.8.1 Preparation to the Technological Systems' Design

This preparation includes:

- development of the physical basis of the projecting;
- development of methods for NDT of materials, workpieces, and elements of control systems;
- qualimetry data investigation (on construction analogs);
- calculation methods and mathematical models development;
- analysis of modeling results.

21.8.2 Preparation to Diagnostics in the Design of Technological Systems

This preparation includes:

- investigation of technological processes' regimes;
- investigation of technological equipment, and elimination of defects;
- inclusion of qualimetry data and oscillograms in technical documentation.

21.8.3 Investigation of the Technological Equipment at the Stage of its Assembly

This investigation includes:

- testing in assembly;
- investigation of shaking-down period length (unique for every unit);
- definition of the qualimetry data and standard oscillograms;
- certification of the new construction;
- audit of the plant.

21.8.4
Investigation of the Technological Systems at the Stage of Operation

This investigation includes:

- testing materials, instruments, workpieces, and control systems;
- testing of the technological systems;
- periodical diagnostics of technological equipment and instruments;
- testing of production;
- diagnostics before and after repair and modernization;
- certification and audit of the plant.

21.8.5
Evaluation of Data and Knowledge Bases

Data and knowledge bases include:

- quality indices' nomenclature;
- the results of NDT;
- qualimetry data on mechanisms;
- typical faults and fault trees on mechanisms and technological systems;
- rules and procedures; selection of a method for quality indices' calculation.

21.8.6
Personnel Education, Instruction, Training and Certification

These are the leading tasks of increasing productivity, quality of production, trouble-shooting, and safety of technological systems.

21.8.7
Evaluation Methods of NDT and Diagnostics and their Application in Safety Management

The main principle of policy in the safety field is strategy of risk reduction and protection of the population and the environment from emergencies (extreme situations).

The connection between safety management technological systems and other diagnostic systems (environmental, medical, and criminogenic) must be investigated.

21.8.7.1 Technogenic Safety

As a rule, to provide the technogenic safety of complicated objects it is necessary to control a wide range of different parameters, such as kinematic (time, velocity, flow, acceleration, etc.), static and dynamic (mass, force, pressure, energy, etc.), mechanical (specific weight, substance quantity, density, etc.), geometrical, electrical, thermal, magnetic, acoustic, and other physical parameters. Nowadays, the role of methods and means for flaw detection, introscopy, structure analysis, dimensional measurements, and monitoring of physical and mechanical parameters of materials and units has been considerably extended.

21.8.7.2 Social and Economical Safety

The NDT methods and means that solve the problems connected with the necessity of finding the arms, explosives, and radioactive and nuclear sources, as well as of detecting false documents and securities are rapidly developing.

The essential part of designed and produced devices and instruments is applied to territories' protection and monitoring, and watching after criminals.

The following Table 21.8 gives a brief presentation of physical methods used for different diagnostic procedures:

At present, all over the world the X-ray television systems, and instruments based on magnetic, thermal, and eddy-current methods are used to check the air-passengers' luggage and to check the staff when entering hazardous areas. In banks, security services and the optic-television and endoscopic technical-vision systems are widely used.

21.8.7.3 Ecological Safety

The methods and means for ecological diagnostics made rapid strides among all the NDT-developing areas. To provide atmospheric monitoring, good results have been achieved recently in the development of surface-acoustic wave sensors (SAW), laser measuring systems, infrared detectors, and systems based on other physical principles.

The other area in the interest of ecological diagnostics is the water, water basins, and ice-condition monitoring. Electromagnetic emissions in the radio wave range, and ultrasonic methods are used for long periods of time for diagnostics of water basins.

Table 21.8 The diagnostic tasks, methods, and means.

The diagnostic tasks	Methods and means
To find arms	Magnetic, electromagnetic, infrared metal detectors, radiographic-television introscopes, fluoroscopes
To find explosives	Gas analyzers, chromatography instruments, drift-spectrometers, neutron flaw detectors, nuclear-magnetic and nuclear-quadruple-resonant instruments
To find the radioactive and nuclide sources	Dosimeters, radiometers, gamma-signaling devices, spectrometers
False documents and securities detection	Ultrasonic, infrared, optical, and X-ray instruments
Territories protection, search, and watch	Optical, ultrasonic, infrared, and TV systems of technical vision

The most extensive information about wood structure changes is obtained with the help of optical and X-ray inspection methods.

A very promising method for ground-bed diagnostics is the radio detection and ranging method, using the same engineering philosophy as the traditional radar method does.

One of the most complex tasks of ecological diagnostics is the forecast of earthquakes. The equipment must provide monitoring over large territories and the possibility to transmit data via satellites to regional diagnostic centers, including a large number of mobile and air laboratories.

21.8.7.4 Medical Diagnostics

The most comprehensive object from the diagnostic point of view still remains the human being with the variety of diagnostic parameters of his different organs, illnesses, etc.

Almost a whole range of electromagnetic-radiation waves, radiation, ultrasound, and electrical fields are used in medical diagnostics as well as a huge amount of different medicines and compounds. At present the average volume of medical diagnostic equipment sales exceeds the volume of NDT means sales. But at the same time the variety of medical equipment is much less than the variety of NDT equipment. More than 80% of all visualized images and more than 50% of material expenses in medical diagnostics are accounted for by X-ray diagnostics.

The common tendency of medical diagnostic means' development is the use of 3-D images and a combination of diagnostic equipment with the means for surgery and therapy.

These complex tasks of diagnostics are very important also for testing and diagnostics of automatic equipment and technological systems. Many instruments are identical and, taking into account the sales volume, they would cost less.

Therefore, the NDT and diagnostic methods should be widely applied for estimation of defects and major characteristics of equipment and technological systems at all stages of design, manufacture, and application.

Literature

[1] Klyuev VV (1998) Non-destructive Testing and Security Diagnostics.J, Industrial Laboratory, N1, pp 66–68

[2] Klyuev VV (1995) Non-destructive Testing and Diagnostics (in Russian). Machinostroenie, Moscow

[3] Klyuev VV, Sosnin FR, Filinov VN (eds) (1996) Measuring, Inspection, Testing and Diagnostics, Mechanical Engineering Encyclopaedia, vol III–7 (in Russian).Machinostroenie, Moscow

[4] Klyuev VV, Gusenkov AP (eds) (1998) Reliability of Machinery, Mechanical Engineering Encyclopaedia, vol IV–3.Machinostroenie, Moscow

[5] Safety (in Russian) (1998) Moscow; Function and Development of Complicated Systems (in Russian) Knowledge

[6] Safety (2000) Environmental Diagnostics (in Russian) Knowledge, Moscow

[7] Nachapetjan EG (1990) Testing and Diagnostics of Automatic Equipment (in Russian). Nauka, Moscow

[8] Gusenkov AP, Nachapetjan EG (1993) Methods and Means for Securing Machine Reliability (in Russian) Nauka, Moscow

[9] Nachapetjan EG (1985) Diagnostics of FMS Equipment (in Russian). Nauka, Moscow

[10] Nachapetjan EG (1979) Qualimetry of Robot and Manipulator Mechanisms. In: Proceedings of the Fifth World Congress of Theory of Machines and Mechanisms, Montreal, Canada, pp 835–838

[11] Bently(1992)Rotating Machinery Information Systems and Services. Monitoring Systems. Nevada

Part V
Future Trends

Chapter 22
Prospects of Technology Development

FANUC LTD

22.1
Technologies for Factory Automation and Robots for Manufacturing of the Future

The 21st century has come with the great impact of information revolution to change our society towards the new creation of the information society with much emphasis on individual pleasure through convergence of time, distance, and place, with instantly networked communication and information processing. Its nature of uncertainty is due to the inexperienced new individual-based communication and network, which may be received with confusion and chaos on the traditional mind; however, it will be transformed into an excellent opportunity of challenge and dynamism for the upcoming generation. Factory automation and robots will perform to merge our traditional industrial society with the new information society. And factory automation and robots with networked intelligence will generate an unlimited opportunity as a bridge to bundle two cultural behaviors in manufacturing. New vision for the future will be provided with our insights on factory automation with information technology.

22.1.1
Historical Overview of Factory Automation Technology

Manufacturing in the 20th century was characterized by the evolution of industrial society focused on efficient mass production to rapid market expansion through competitive mass supply. The factory automation and robot have contributed to convert small lot-size production into a seamless flow of production as if it were mass production.

22.1.1.1
Numerical Control

The first commercial numerical control was released in 1958 by FANUC, which concept was originated with the theoretical idea at MIT. The commercialization was led to success due to two critical unique inventions. One was the electro-hydraulic-pulse motor, which was recognized as an advanced integration of digitalization primarily in analogue period, to convert difficult analogue controls into relatively easier digital open-loop controls with smart economical methods. The other was the pulse-interpolation principle with arithmetic calculation, which prompted the simpler and speedier path-motion generation with simple adders as hardware when no microprocessor was invented yet. Fortunately, the popular commercialization of integrated circuits (IC) did accelerate effective competitiveness of numerical control (NC). It resulted in the epoch-making and highly reliable open-loop numerical control at low cost, and eventually contributed to the worldwide competitive differentiation of Japanese machine tool industries through the IC-based NC in 1966, and through the computerized NC in 1972. The progress was significantly supported by the advanced control theory as well. Such technological product advancement prevailed to significantly popularize NC machine tools as the flagship of automated machining, and consequently success has been realized in the Japanese machine tool industry as one of the most forward-looking industries in the world for machine tools. Now, the installation of FANUC numerical control has reached over 1 million units worldwide. And 94% of all new production of machine tools are with CNC. The significant enhancements for high precision, high speed, and high performance have been recognized.

At that time, Direct Numerical Control (DNC) to connect multiple NC machine tools with a computer system was realized for centralized production scheduling, monitoring, and management. The system was advanced towards flexible manufacturing systems (FMS) and integrated manufacturing systems (IMS) with the distributed networked system, and now as the intelligent manufacturing system as an extended concept.

22.1.1.2
Industrial Robot

In the same period of time, the original invention of the hydraulic industrial robot was commercialized in the US for mass production, focusing on the automotive industry. Then, the CNC technology with electric servo motor was introduced in the industrial robot, and it significantly contributed to enhance the reliability and cost performance of the robot.

The industrial robot, due to its generic motion, has prevailed for manufacturing-process applications. The industrial robot became popular in the automotive industry for spot welding, painting, laser processing, arc welding, sealing, material handling, and assembly in the 1980s, then in non-automotive industries in the 1990s. The worldwide robot population is now over 700,000 units in active operations for manufacturing, 68% of which are in Japan.

Now, let's focus on the role and contribution of the factory automation with robots for manufacturing purposes.

22.1.2
Technological Advancement Towards Intelligence

A robot has as its mechanism a flexible muscle with its controller as an intelligent brain. The robot is conceptually universal and flexible, and its flexibility has been expanded through integration of dedicated skills and know-how for specific process solutions into the robot controller, with versatile motion.

For example, in the robotic paint process, the stability of spray finish at surface contact as well as the smoothness at the tool center point have provided excellent flexibility and feasibility to fit closer into paint process environment. Two vital directions are recognized for flexibility in motion and specialization of process know-how for intelligence.

Reviewing an articulated arm as a most typical configuration of the industrial robot, its mechanism consists of base 3 axes for position and additional wrist axes for orientation. Each of these axes is controlled individually with spring and dumping factors as a control model. Dynamic variation of load and inertia condition can be adjusted by advanced control theory with dynamic optimization such as sliding mode control, gain scheduling, torque feed-forward control and so on, for enhancing rigidity and safety.

Recognizing robot intelligence and its sophistication, visual sensing and force sensing contribute to important instant status of external environment into robot control, and its software knowledge base contributes to recognizing the dynamic environmental conditions and to optimizing robotic behavior, such as avoiding possible risks. The intelligent robot with process solution plays a key role in the 21st century for manufacturing.

22.1.3
Market Globalization

We are in a new era with information technology. The convergence of computer and communication networks provides instant individual communication anytime, anywhere, and worldwide. It merges together information flow with materials flow in our manufacturing. It generates significant changes in production efficiency beyond our mass production in the industrial society. Instant global communication transforms individual local tastes to worldwide global tastes instantly in our market.

Such trends create new opportunities and challenges in competitive production worldwide. De facto standards are much more favorable for global conformance (see Chap. 2). Decentralized and distributed information and communication flow provide flexibility for seamless upgrades of system functionality. Much opportunity has come for individual leadership and the small-enterprise spirit. Large establishments are hustling to accommodate new trends into their established organization by reforming its organizational structure, due to their decision and authority system. For example, the integration of information technology brought significant efficiency into production in manufacturing. It resulted in excess capacity in production resources as well as in human resources. The ownership of goods became less important, and much emphasis is now on control of most information to challenge the future.

In the traditional society, smart skills and talents have been enhanced through expertise with seniority. Expertise is the integral part of human resource management. It gives superiority favor with complimentary competition among larger enterprises within the same industrial sector. Digital economy in the information society has put emphasis and value more on information than goods. It is also favorable for individual inspiration beyond expertise with a youthful spirit and opportunity for challenges.

Loosely networked intelligent robots provide new flexible production environments with integration of information and materials flow, beyond quantity focus on conventional wisdom in the industrial society. De facto standards are important for such flexible connections among cross-industrial societies.

22.1.4 Challenges for Intelligence with Flexible Automation and Robots

An intelligent robot work-cell is an application unit for certain performance objectives. It automates the recognition of parts location and environment. It integrates skilled know-how. It works out with appropriate optimal judgment. It watches any unusual situation and alarms or stops its sequence. For optimal performance, it uses countermeasures through recognizing the working environment. Such features are currently available. In addition, we are expecting much superior intelligence to interact the robotic cell with the human workforce.

In robot technology and commercial product development, vertical organizational integration was well performed as key elements in the past. Now with such intelligence, we do need transcendental cross-functional activities with synergy among various technologies as well as among various industries. The versatile cooperation is becoming significantly more essential than it used to be.

Reviewing global constraints on the natural environment, sustainable manufacturing is a key philosophy in the 21st century to minimize any negative environmental impacts to the earth. This situation in the environment on earth has also required the direction mentioned.

Some national projects support the direction for such intelligence with excellent collaboration of research institutions and industrial enterprises with governmental sponsorship. The Humanoid Robotics Project (HRP) is an example of such an approach as one of the Japanese national projects to focus on remote communication with robots and humans. The Intelligent Manufacturing System (IMS) program features the industry-driven international collaboration with governmental sponsorship, initiated by the Japanese government.

The technology for intelligent robots has progressed beyond industrial robots for manufacturing, and is expected to find new value in service and personal robot areas. The success in intelligent automation and robots depends on process know-how in the robotic system by organizing human-oriented expertise and know-how into a technical structure separate from human talents. Such an approach is significantly critical to bring the intelligent robot closer to human behavior. The significant support of such directions is given in the Digital Meister Project, sponsored by the Japanese government, starting in 2001 for 5 years.

22.2 Factory Automation Technologies

22.2.1 Control Technology for Machine Tools (CNC)

22.2.1.1 CNC Hardware

Enhanced machining productivity is one of the major demands being made on the latest models of machine tools. To satisfy this demand, many technologies such as high-speed positioning, high-speed and high-precision machining, multi-path machining, milling/turning-mixed machining, and so on have been introduced to the latest machine tools. However, such technologies can be realized only with a CNC that offers excellent control capabilities. Other demands being made of machine tools are smaller footprints and, of course, lower prices. Recently, the compact lathes that are used for machining small components needed for IT applications have been in popular demand. Also, compact machining centers and compact lathes with a smaller footprint have made an appearance. If a machine tool is to have a smaller footprint, then it is essential for the CNC to be made much smaller. Simplified wiring of the CNC helps to save engineering costs of machine tools.

The latest CNC employs multiple microprocessors, each of which is dedicated to specific functionality, such as CNC control, servo control, sequence control, display control, network control, and so on. The use of multiple microprocessors enables the CNC to provide excellent sophisticated control. The latest CNC hardware consists of a high-speed microprocessor, large-capacity semiconductor memory, Application-Specific IC (ASIC) designed for dedicated functionality, compact connectors, and so on, which ensure the printed circuit board is as small as possible. The package of the ASIC is a so-called fine-pitch Ball Grid Array (BGA) with 0.8-mm pitch pins. A Liquid Crystal Display (LCD) is employed as the CNC's display device. Nowadays, 10.4" color Thin-Film Transistor (TFT) LCDs are extremely popular in machine-tool applications because the dimensions of the LCD are ideal for use in the machine operator panels of the machine tool. Each ultra-compact CNC actually integrates the ultra-small control printed circuit board into the LCD display. The CNC is only 60mm in depth. Such ultra-compact and ultra-thin CNC is ideal for simpler engineering with the power magnetic cabinet.

Reducing cables and wiring is another key factor to reduce the cost of a machine tool. Conventional cabling between the CNC with multiple servo amplifiers requires metal wiring for each servo amplifier. The latest CNC allows a single high-speed serial connection with the plastic optical fiber to connect all the servo amplifiers and the CNC instantly at inexpensive than metal wiring. A special Error Correction Code (ECC) is applied to the high-speed serial connection to assure the validity of the transfer data. Transfer data with an ECC is sent together as a package, and is verified by the ECC at receiving. The ECC ensures most is error-free.

Since versatile devices are utilized in a machine tool, simpler cable and wiring with these devices is also another key important factor to reduce the cost. Distributed I/O devices in a machine tool provide better solutions than centralized I/O devices. The Actuator Sensor Interface (AS-Interface) standards recently established contribute to realizing the distributed I/O devices, such as solenoid valves, magnetic contacts, limit switches, and proximity switches with simpler wiring.

22.2.1.2
High-Speed, High-Precision Machining

The most essential objective of the CNC is to achieve high speed and high precision machining at the same time. However, higher speed machining tends to result in poor precision accuracy, and higher precision machining tends to require slower speed. The latest servo control technology conquers the contradiction, and achieves high-speed and high-precision machining at the same time.

The tolerance in contouring, or the deviation of the actual tool path from the programmed tool path, is primarily generated as the accumulation of the following:

- Delay due to acceleration/deceleration after interpolation
- Delay due to servo lag

Technical solution to reduce such errors incorporated in the CNC is reviewed here:

1. Acceleration/deceleration before interpolation: The tangential feed-rate is accelerated/decelerated, before the path is interpolated with the specified feed-rate, to compensate for any error.
2. Feed-rate verification on each axis at a corner: The feed-rate is verified on each axis at a corner and, any feed-rate exceeding the speed limit is decelerated before interpolation,- so as to minimize possible shock.
3. Feed-rate verification at acceleration: On consecutive small lines commanded to form a curve, the acceleration normal to the tangential direction is verified and the feed-rate is decelerated before interpolation, to minimize any shock.
4. Acceleration/deceleration adjusted before interpolation over multiple blocks: The actual feed-rate for consecutive multiple blocks is determined through the methods described in (2) and (3).
5. Nano interpolation: The nano interpolation means internal interpolation of the programmed path in very minute units or in nano orser, specifically 1/1000 of a command unit, normally of 1 micron-meter. The nano interpolation provides the smooth contouring:
   a Smoother interporation with up-to-date feed-forward control at servo system provides interpolation data directly added to the velocity data, to reduce the error generated by servo lag. The nano interpolation in nano-meters provides a smoother velocity command, to generate a smoother surface on a work piece.

b The existing programs for CNC can be utilized with a command unit of 1 micron-meter.

22.2.1.3 *Five-Axis Machining Features*

Five-axis machine tools with a rotary tool head or rotary table, are popular for complex machining of automobile and aircraft parts. The up-to-date CNC provides prime features to support 5-axis machining:

1. Three-dimensional coordinate-system conversion: The coordinate system is available to rotate around the specified axis with the center of rotation, the direction of the axis, and the rotation angle.
2. Three-dimensional cutter compensation: Cutter-radius compensation is applicable on the plane, perpendicular to the tool direction.
3. Tool center point (TCP) control: The cutter path and its speed at the tool center point can be controlled.
4. Three-dimensional handle feed: Handle feed along the tool direction, in the lateral direction, or around the tool center point are available.

22.2.1.4 *Compound Machining Features*

The popularity of compound machine tools to perform both turning and milling features at the single machine has recently been significant in automobiles and aircraft. The up-to-date CNC specifies the prime features to support compound machine tools:

1. Compound machining: Functionality for turning and for milling is provided with a single CNC.
2. Interference verification for rotation: An interference is verified for a swivel head or a rotary table.
3. High-precision contour control for turning center: High-precision contour control is available at a CNC for turning.
4. Turning functions for machining center: Turning functions such as tool nose-radius compensation, multiple repetitive cycles for lathe, and so on, are available at a CNC for a machining center.

22.2.1.5
Remote Diagnosis and Troubleshooting Guides

In recent years, there have been significant demands in the automotive and aircraft to reduce the Mean Time to Repair (MTTR). The latest CNC provides functions to satisfy the requirement. One solution is a remote diagnostic function to verify the machine and to diagnose the CNC and PMC statuses via a network. The other solution is a troubleshooting guide to track down a failed unit.

1. Remote diagnosis: If a machine tool breaks down, the operator can check the CNC/PMC status remotely over a network. The following operations are supported:
   a Remote PMC diagnosis and editing
   b PMC signal status / PMC signal trace / PMC parameter, etc.
   c Remote operation of basic CNC screens, such as Position / Offset / NC program / Macro variables / NC parameter / Servo spindle waveform diagnosis data / Operation and alarm history.
2. Troubleshooting guides: Possible causes and questions are displayed for troubleshooting with an alarm. Selection of “YES” or “NO” on the screen leads to the appropriate solution in simple operation to track down the actual cause of a CNC alarm.

22.2.1.6
High-Speed PMC

The Programmable Machine Controller (PMC) is built into the CNC, to provide sequence control of various functions for a machine tool and its peripheral operations such as automatic tool changer, palette changer, and parts loader. The PMC-has a dedicated microprocessor and extended program memory independent from the CNC for high-speed and high-efficiency machining through its high-performance sequence control. The PMC is also capable of diagnosing and editing a sequence program with the CNC display without any assistance of external devices to save costs on devices as well as machine operation with mimimum downtime. Along with the networked devices for factory automation in rapid progress on Information Technology, the PMC contributes to the interoperability among the control equipment, operator interface and monitoring devices through factory automation.

22.2.1.7
Open CNC

CNC machine tool demonstrates its full ability when efficient operation is realized in addition to high-speed and high-precision machining in production. Efficient operation in production is assisted with several considerations:

- Human-machine interface for simpler and easier operation.
- Programming and program database to fit individual configuration of machine tool.
- Simplified interface to connect machine tool and other devices in production.

An effective alternative to satisfy such requirements is to integrate the PC with its software in a CNC. The CNC with integrated PC is called an Open CNC, which expands PC applications to enhance functionality for supervising management as well as program support.
Figure 22.1 shows the configuration of the Open CNC. The PC and CNC/PMC are connected via the FANUC High- Speed Serial Bus, a serial communication over the Ethernet LAN with optical fiber. The PC with the Windows operating system enables one to utilize software commercially available, such as CAD/CAM, databases, and networking. Data communication between PC and CNC/PMC is supported by the BASIC API functions with the interface such as "FOCAS1" i.e., FANUC Open CNC API Specifications, version 1, which includes more than 300 functions, to ac cess and process almost all the essential data of CNC/PMC. The FOCAS1

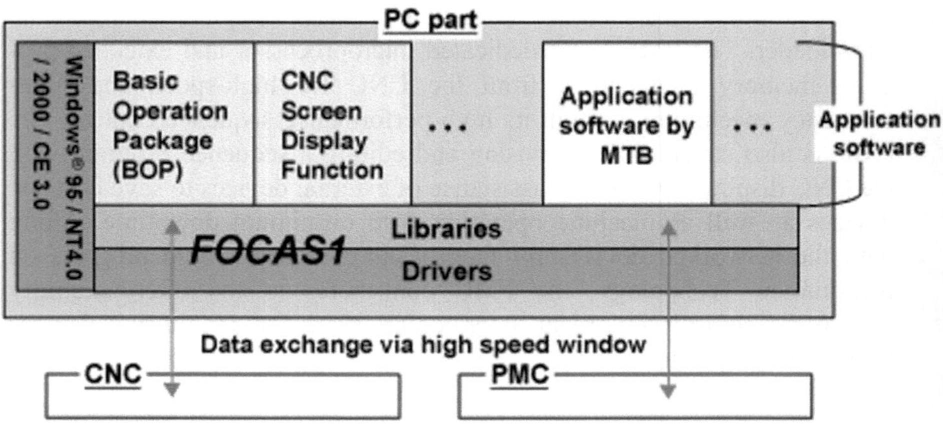

Fig. 22.1 FANUC open CNC configuration.

interface is standardized with CNC/PMC, so that customer application can be easily ported among Open CNCs.

A specific Open CNC with the Windows CE operating system, with no requirement of a hard disk as storage, can realize high reliability in the same way as the CNC without a PC.

22.2.1.8
Features of a Flexible Transfer Line

Transfer line for the automotive industry requires more flexibility than in the past. A wide range of versatile CNCs, peripherals and displays, and a network for factory automation are available to meet customer expectations on flexible transfer lines.

The touch-screen is one of the key devices to enhance flexibility on transfer lines, by configuring the operator panel only with essential buttons for easy and intuitive operation to allow limited selections. In the past, re-configuration of production lines required its redesign to reflect the new configuration. The touch-screen allows one to create new requirements easily through software modification. A single display can be shared with multiple CNCs, to display information for individual machines including maintenance information for diagnosis.

22.2.2
FA Network

22.2.2.1
Overview of FA Network

The FA network is essential in flexible automation to communicate between CNC machine tools and devices in production. The various styles of communication with the network are required, though the communication is generally classifiesd in three hierarchies (Fig. 22.2).

1. Area network: The area network is a link at the entire factory level with communication among the central supervising computer and cells on the shop floor. The Ethernet (TCP/IP) protocol, a global standard in office automation, is generally used for the area network applications.
2. Cell network: A cell is defined as a system consisting of multiple independent CNC machine tools. Communication in the cell is performed under the controller in the cell. The Ethernet protocol is also in popular demand for this level of control.

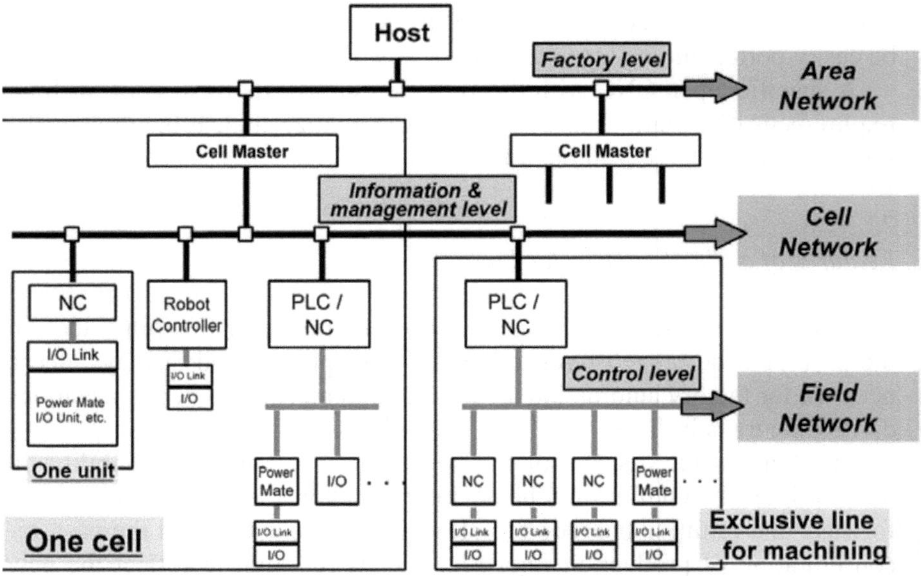

Fig. 22.2 Hierarchies of FA network.

3. Field network: The field network is a link to communicate peripherals and I/O devices with CNC machine tools. The programmable logic controller (PLC) is available to supervise the control of all the CNC machine tools and conveyors of handling workpieces. Such communication is supported with high-speed transfer of control signals among devices and peripherals.

22.2.2.2 Cell Network

The cell is controlled through signals such as input and output of CNC programs, control for operation, and so on. The cell is supervised with operation status and is managed within a schedule. The Internet technology with Ethernet makes the system simpler in its network (Fig. 22.3). The collected information at the factory is shared beyond the system and enhances production efficiency with the overall floor communication. A supervisor may easily monitor the operations from a remote location through a mobile device with Internet access. Moreover, it provides diagnosis and support in any emergency events or malfunctions for the CNC and machine tool supplier over the Internet. This is extremely effective–for quick-fixing to reduce downtime with support and service on technical and parts information, including how to use. Widespread sharing of such accumulated information makes further value for the future.

As described above, the application with network-to-factory automation is significantly important to cooperate with computers over the Internet.

22.2.2.3
Field Network

A field network for machining requires high-speed response with real-time performance to control multiple devices. The network should be standardized to link various devices among different vendors. Standard protocol such as Profibus, DeviceNet, and so on are now popular in factory automation.

Moreover, the field network is also applied to the device level such as I/O devices, sensors, etc. The requirements on this level are different from the cell level on its controllability. For example, a ladder program is executed through the controller to oversee the entire line, and production lines are deeply dependent on such information. There is a trend toward the Ethernet away from field networks.

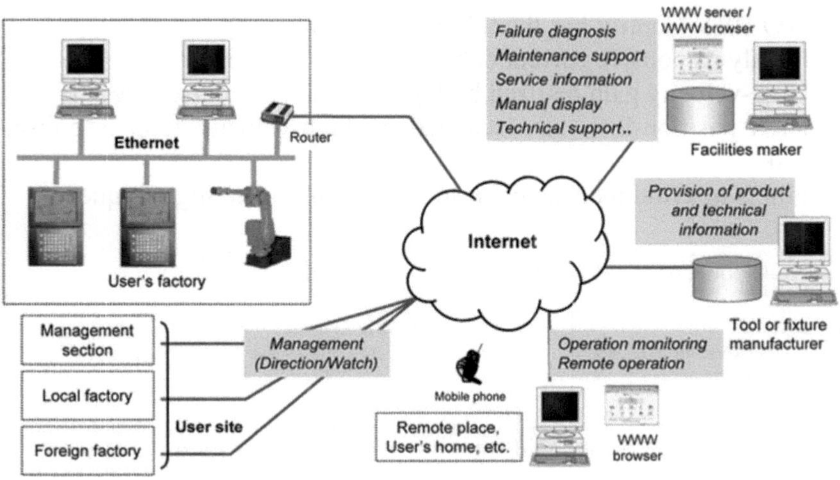

Fig. 22.3 Internet technology applied to FA.

22.2.3
Servo System

22.2.3.1
High-Speed and High-Precision Servo System

1. Technologies required for high-speed and high-precision servo systems: As shown in Fig. 22.4, "High-speed and high-precision response to input command" and "Robustness against external disturbance such as friction or cutting torque" are required to realize a high-speed and high-precision servo system. To this end, a "CNC (command generator)," "Servo Motor (actuator)," "Servo Amplifier (power amplifier)," "Pulsecoder (position and velocity detection)" and "Servo Software (Control)" are required to produce a well-balanced combination.
2. SERVO HRV Control – High gain and feed-forward control: Among the five technologies above, we would like to discuss a Servo Control technology called SERVO HRV Control, which realizes a responsive, high-gain servo system. Servo control has three control loops, namely, current control, velocity control, and position control loops as viewed from within.

   SERVO HRV Control has the following features:
   - Highly responsive current control by high-speed DSP,
   - High-velocity gain realized by highly responsive current control,
   - High position-loop gain realized by high-velocity loop response, and
   - Filters to avoid mechanical resonance from low to high frequency.

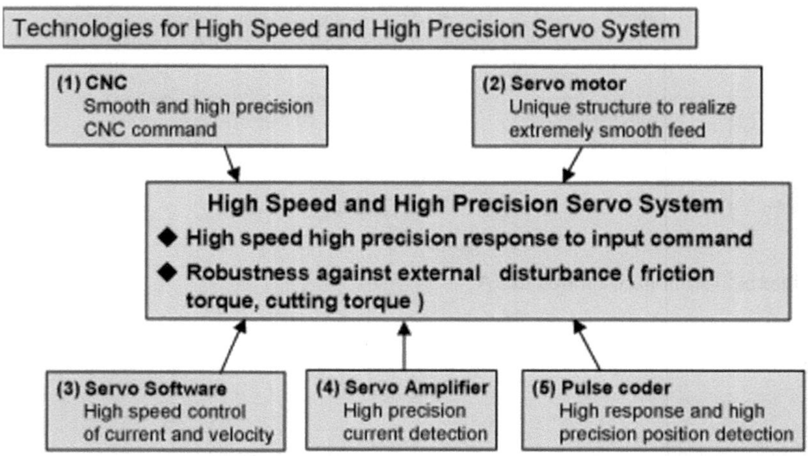

Fig. 22.4 Structure of the system.

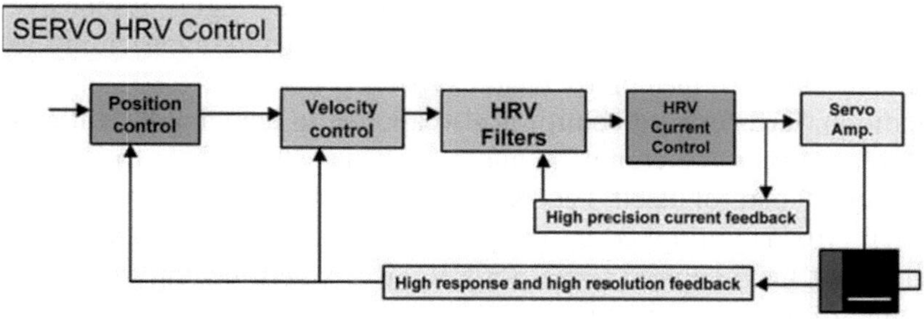

Fig. 22.5 Major factors of the control characteristics.

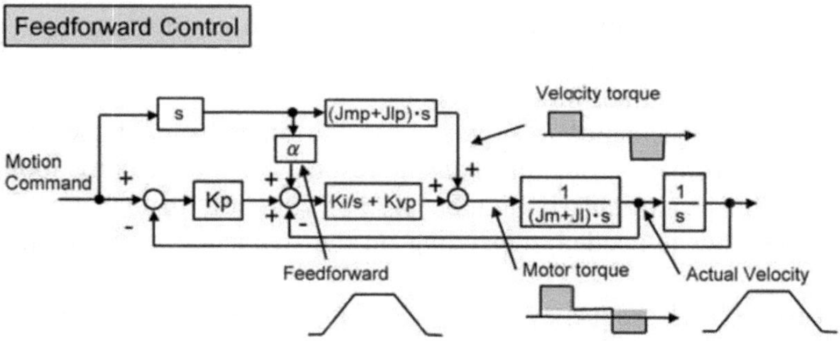

Fig. 22.6 Scheme of the feed-forward control.

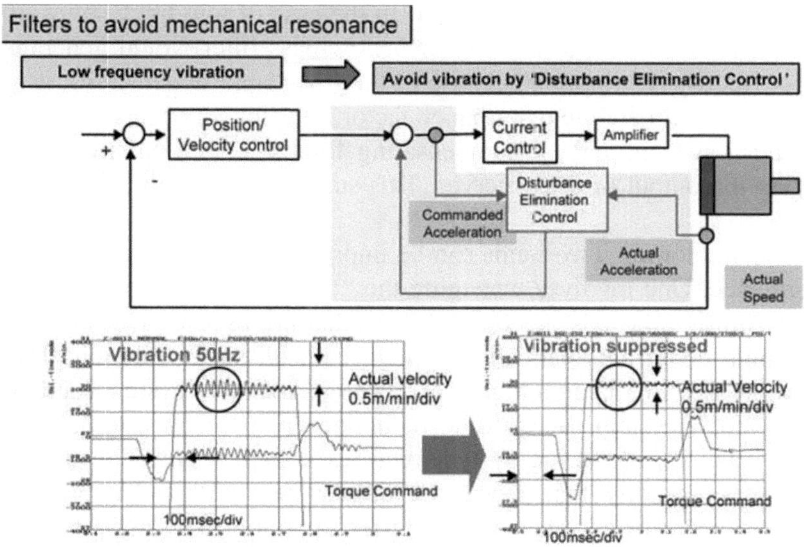

Fig. 22.7 Scheme of the device to suppress mechanical resonance.

Of course, “high S/N ratio current detection” and a “highly responsive and high resolution pulsecoder” are also major factors contributing to the enhancement of the control characteristics (Fig. 22.5).

In addition, feed-forward control is also applied to reducing the following error between the command and the actual path. This is caused by the position control itself generating a velocity command when the position error is multiplied by the position loop gain. This means that, “there must inevitably be some position error when feeding the axis,” and this following error causes a path error when there is an acceleration of the command in the radial direction. So, in order to reduce this error without setting a position loop gain with stability, feed-forward control should be applied (Fig. 22.6). This feed-forward control reduces the following error while a high-gain servo system corrects an unexpected following error caused by disturbance torque. Together, these realize high-speed and high-precision control.

3. Filters to avoid mechanical resonance: On the other hand, if the machine suffers from a strong mechanical resonance, the velocity loop gain cannot be set particularly high because of vibration. In such a case, an adequate filter is needed, depending on the frequency of the vibration.

   In the case of high-frequency vibration of more than 200Hz or so, a notch filter that has a sharp-decrement characteristic is effective. The vibration frequency and bandwidth frequency of the velocity loop is so far away that the phase delay of the notch filter does not have a secondary effect on the servo control itself. In the case of low frequencies of less than 100Hz or so, a notch filter or low-pass filter would generate another frequency vibration because of the filter’s phase delay. So, it is necessary to estimate the disturbance factor by using an observer inside the control system, and by compensating for the torque command by applying the output of the observer. This suppresses the vibration (Fig. 22.7).
4. Future technologies: Two items can be improved to obtain higher servo performance. One involves managing the “‘time-drifting resonance depending on time or axis position.” Filters that are used to avoid vibration should adapt automatically when faced with such vibration. The second involves realizing a higher precision direct-drive system such as a linear-motor system. Because no reduction or disturbance torque directory acts, a direct-drive system requires a higher precision and a higher servo rigidity than a rotary-motor system with a ball screw. There is a need for a system with a faster sampling time, as well as a more sophisticated algorithm to avoid resonance.

22.2.3.2
Servo Tuning Tool

1. Necessity of the Servo tuning tool: Servo/spindle systems are being developed with better and better performance and, accordingly, the systems and their tuning are becoming more complicated. For a machine-tool builder, it becomes difficult to extract the full performance of a servo/spindle system while fitting each machine's characteristic. Suitable servo-tuning tools are becoming necessary for both optimum and quick tuning. So, FANUC developed SERVO GUIDE.
2. Tuning procedure: The system configuration includes three windows:
   - Parameter Window: Displays the servo and spindle parameters for each function.
   - Program Window: Automatically generates the CNC programs for servo tuning.
   - Graph Window: Offers methods for displaying and analyzing actual movement.

   Servo/spindle-related parameters and tuning programs are output to the CNC and executed. The result is shown in the Graph Window. Figure 22.8 shows examples of the Graph Window. The optimum servo/-spindle parameters can be obtained by repeating this process several times.
3. Future development: Of course, an auto-tuning feature would be extremely helpful. From a system-configuration viewpoint, simply adding an "Auto-Tuning Window" which "reads the result from Graph Window," "considers how to tune parameters to reach the desired result," and which "changes parameters through the Program Window" would realize an auto-tuning system. From a software viewpoint, a "Tuning Engine," an expert system that depends on servo theory and experience of the actual machine, will be constructed in the near future.

22.2.3.3
Energy-Saving Servo System

1. Energy-saving technologies: We continue to develop high-speed and high-precision servo systems and also energy-saving servo systems. Here, we introduce the results we have achieved with our energy-saving servo system. Our approach to realizing energy saving in a machine tool is done from two aspects:

   One involves the "direct reduction of energy consumption" (DREC) while the other is the "indirect reduction of energy consumption (IREC)."

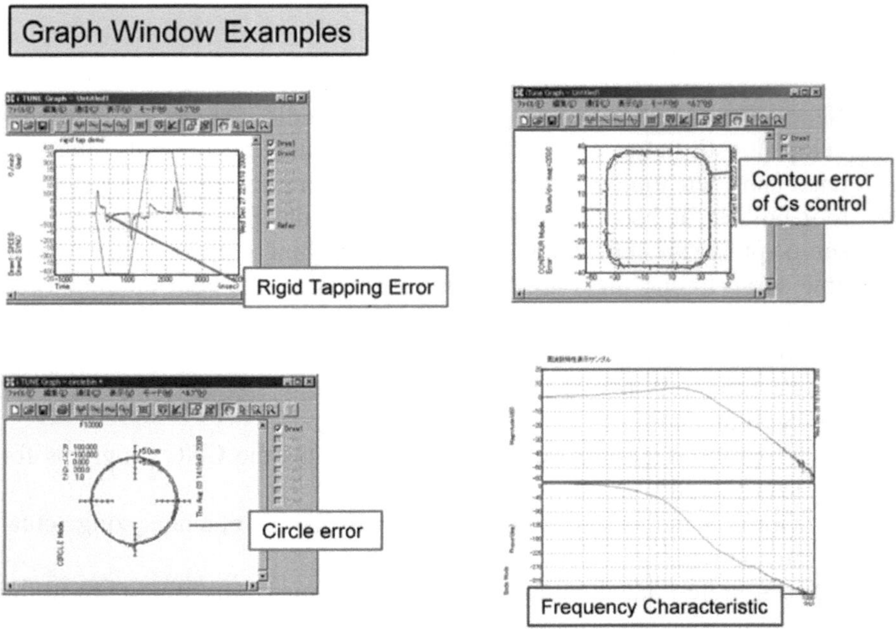

Fig. 22.8 Examples of results.

The DREC involves reducing energy consumption by improving the efficiency of the servo system itself.

The IREC involves reducing the energy consumption of the fan motor, hydraulic pump, etc. by increasing the speed and acceleration of the servo system and hence shortening the cycle time.

The main technologies that support energy saving are listed in Table 22.1.

The following explains the three technologies for enabling the DREC, as shown in Fig. 22.9.

a Power-source regeneration: By adding our original phase-firing control to power-source regeneration, we were able to realize highly efficient power-source regeneration where distortion of the regeneration current becomes small and the power factor is improved.

b Power-loss reduction through use of the latest power devices: We have low-loss power devices that use state-of-the-art process technology as the main switching transistors. And, by developing a specific transistor for converter use in which a lower loss rectifier diode is installed, the loss in the transistor can be reduced by about 20% to 30%.

Table 22.1 Main technologies that support energy saving.

Direct reduction of energy consumption	Indirect reduction of energy consumption
1) Power source regeneration	4) Reduction of cutting time by application of SERVO HRV Control
2) Power loss reduction by using latest power devices	5) Reduction of acceleration time by application of SPNDLE HRV Control
3) Spindle-motor loss reduction by maximizing efficiency control	6) Reduction of acceleration time by adoption of aM Servo Motor
	7) High-efficiency cutting by adoption of aT Spindle Motor

Table 22.2 High-speed, high-precision, and high-efficiency energy-saving servo system.

	Conventional servo system	New Energy-saving servo system	Saving by %
Cutting time	89 seconds	28 seconds	-69%
Integrated watt per one cutting cycle	23.8Wh	10.7Wh	-55%

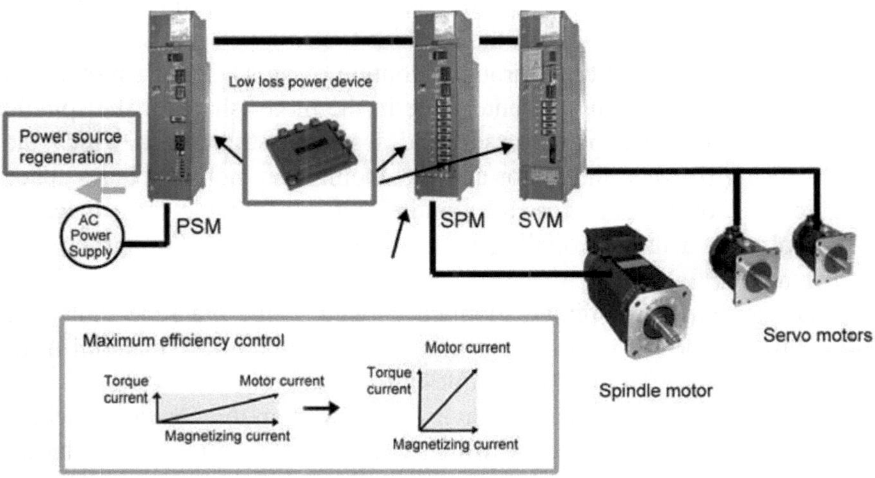

Fig. 22.9 Structure of the DREC.

c Development of maximum efficiency control (Spindle Motor current control): Current control for the induction motor basically involves keeping the magnetizing current constant regardless of the load. We have applied a new current-control method to keep the relationship between the load current and the magnetizing current equal (1:1) at light loads and to realize the most efficient current control for the induction motor (Spindle Motor).

d Reduced cutting time through adoption of SERVO HRV Control: We have added the following technologies to the d-q axis-conversion vector control:

  i. Voltage-clamping method to stabilize the current control when the input voltage is low.
  ii. Improved acceleration ability by making the terminal voltage of the motor wiring lower by optimizing the current phase at high speeds.

  By applying these (SERVO HRV Control) methods, high-speed and high-precision servo control is realized while maintaining the same accuracy of circular cutting while reducing the cutting time in half.

e Reduction of acceleration time by adoption of SPNDLE HRV control: To utilize the motor-applied voltage with maximum efficiency, we added compensation data according to the center value of the three-phase voltage command to the Pulse Width Modulation (PWM) control. The motor torque is improved and the acceleration time is reduced by about 12% in the case of a 5.5-rated output-spindle motor.

f Reduction of acceleration time by adoption of an aM servo motor: By developing an original magnetic pole structure in which a neodymium magnet is arranged radially, an aM servo motor realizes high-acceleration characteristics with low inertia and high torque, thus maintaining a very smooth axis feed of the machine tools.

g High-efficiency cutting through adoption of an aT spindle motor: An aT spindle motor has a center hole in the motor shaft for the spindle through which coolant passes, and a special structure in which the housing between the motor and the cooling fan motor includes space for the rotary joint. This aT spindle motor is suitable for a machine tool with a direct coupling to the spindle with the through-coolant function.

  By applying the spindle-supplied coolant to the machine tools, the cutting time for deep-hole drilling and high-hardness material milling is reduced to one-half or one-third.

2. Effect: We applied this energy-saving servo system to a drilling machine and confirmed the degree by which energy consumption was reduced. The cutting time is reduced to about one-third and the energy consumption is reduced to about half.

We have realized a very high-speed, high-precision, and high-efficiency energy-saving servo system. We will continue to polish our servo system further, not only by "direct reduction of energy consumption" but also by "indirect reduction of energy consumption."

22.3 Ultra-Precision Nano-Machine Tool

22.3.1 Ultra-Precision Micro-Mechanical Machining

To date, micro parts used in the Micro-Electro-Mechanical-System (MEMS) field have usually been fabricated by silicon-based micro-machining methods such as photolithography and etching technology. However, mechanical machining should be re-examined, because it has several advantages over the silicon-based technology. For example, it is comparatively easy to machine sloped and curved surfaces. Moreover, it is easy to create sharp edges and smooth surfaces, and to select the material to some degree. It is considered that mechanical machining is not suitable for volume production. However, it is possible to solve this problem when molding with plastic, glass, and so on, by using a highly accurate metal mold to create the product. At FANUC, we have been researching the potential of mechanical methods for micro-machining. In this section, we would like to focus on samples of ultra-precision micro parts, machined by our ultra-precision machine, the FANUC ROBOnano Ui.

Mechanical machining includes cutting, grinding, and Electro-Discharge-Machine (EDM). In this section, we focus on cutting methods that use a single-crystal diamond tool as a way of improving the form accuracy and the surface roughness. There are two variations of the cutting technology. The first involves making the tool with a high degree of accuracy and then accurately transcribing it onto the workpiece. This method can shorten

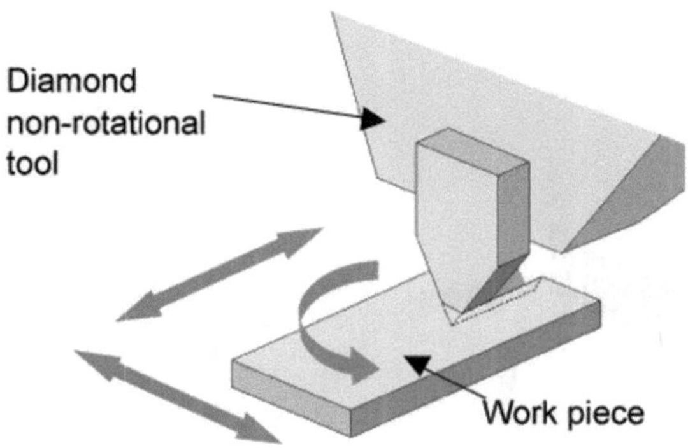

Fig. 22.10 Scheme of non-traditional diamond tool.

the total machining time, and it is easy to create the machining program. However, there is a limit to the form accuracy of a tool. The second method involves tracing machining shapes with a pseudo ball-end-mill with a micro R-shaped tool tip. This method can correspond to complex shapes, but the total machining time is long. The method used depends on the shape to be machined. On the other hand, there are two types of tools that can be used, namely rotation and non-rotation types. Rotational tools are not suitable for machining curved grooves, because of the interference that occurs at the beginning and the end of the cutting. However, it can easily be machined using a non-rotational tool (Fig. 22.10).

22.3.2 FANUC ROBOnano Ui

Figure 22.11 shows the "FANUC ROBOnanoUi." This machine is a five-axis machining center with three straight-line axes (X, Z, and Y) and two rotational axes (B, C). The resolution of the CNC X, Y, and Z axes are 1nm, while the feedback unit is 1/3 nm. The resolution of both the B and C axes is 1/100,000 degree. And, this machine has an air-turbine spindle for performing high-speed micro milling (max 100,000 rpm) with minimum

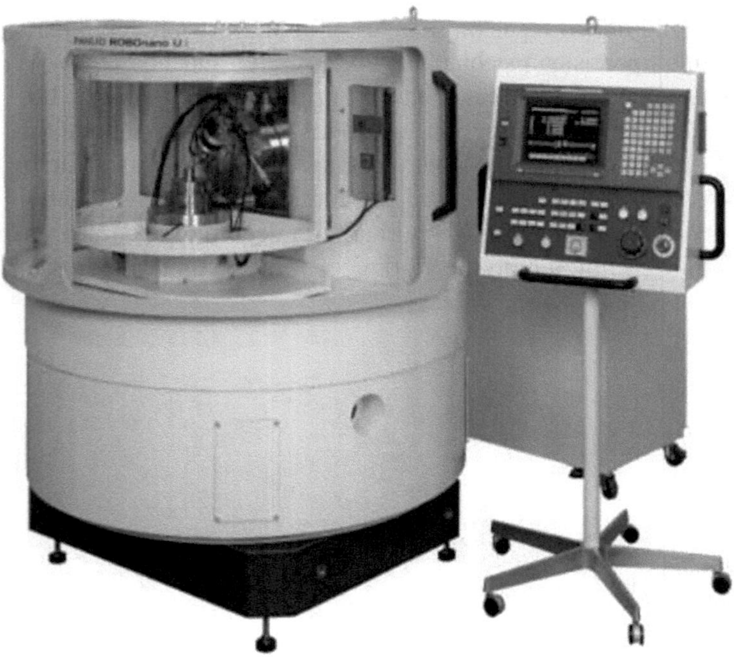

Fig. 22.11 FANUC ROBOnanoUi.

vibration. The main feature of the ROBOnanoUi is that full aero-static air bearings are adopted in all the moving sections (slides, feed screws, nuts, and all motor units), so that solid friction is totally eliminated. For nanometer positioning, overcoming the very small stick-slip-motion caused by friction is the biggest obstacle. Therefore, if a very minute current is theoretically applied to the motor, the slide must produce a correspondingly minute movement.

22.3.3
Examples of Ultra-Precision Micro-Grooves

Many examples of micro-machining, such as ultra-precision micro-grooves like those used for optical lenses, mirrors, filters, and diffraction gratings, are expected to find widespread applications.

22.3.3.1
Micro-Grooves with High-Density Pitch

Figure 22.12 shows V-grooves with a 1-μm pitch and a V-angle on the nickel phosphorus plated steel of 140°. Although these grooves seem to be straight and parallel, this picture shows part of 32,738 radial grooves on a diameter with a 10-mm circumference. We can see that each edge of the grooves is sharp with no burring and each surface is smooth. Both form accuracy and surface roughness are in the nanometer order. In this case, to machine straight grooves, we often use milling with a single-crystal diamond rotational tool. The tip shape corresponds to the V-groove shape. In

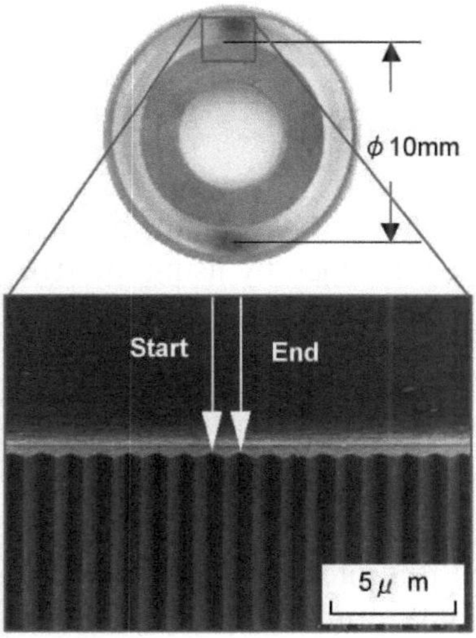

Fig. 22.12 V-grooves with a 1-μm pitch.

the illustrated case, it took 6.5 h to complete the job, and the difference between the depths of the first and last grooves is 30 nm. This may be due to the small amount of heat that the machine generates and the temperature control of the machining environment being within ±0.1 °C.

22.3.3.2
Micro Grooves with a High Aspect Ratio

We have a sample of trapezoid grooves with a high-aspect ratio. The V-angle is 3°, the groove pitch is 35 μm, the depth is 100 μm and the flat part is 15 μm. It was also machined using the diamond rotational tool. It is possible to machine such a structure if a long and slender micro-size tool can be made. These features of micro-grooves with high aspect ratios are common in the manufacturing of semiconductors. However, the roughness of the grooves' sidewalls and the sharpness of their bottom edges are much better than would be possible with semiconductor processing.

22.3.3.3
Free-Shaped Micro-Grooves

A test sample of a mold was realized for a double-focused lens consisting of continuous curved V-grooves of about 35 μm wide with a changeable V-angle and depth. The work material is oxygen-free copper. This was machined by shaping with a non-rotational tool under five-axis control. Although the cutting speed is very low, at 20 mm/min, the machining produced no burring or vibration at the groove edge. Moreover, the intersection of the two lenses exhibited an accurate fit.

22.3.4
Example of Micro 3-D Structure

Not only simple shapes like continuous V-grooves on a plane but also three-dimensional free-curved surfaces can be machined by using the five-axis control function of the ROBOnanoUi.

22.3.4.1
Micro Lens Array

Figure 22.13 shows a 4×4 convex-lens array. Either brass or silicon can be used for this application. The lens has a 290 μm pitch, a diameter of 236 μm, is 16 μm in height, and has a radius of 448 μm. The tool is a single-crystal diamond end-mill with a circular arc shape, which corresponds to

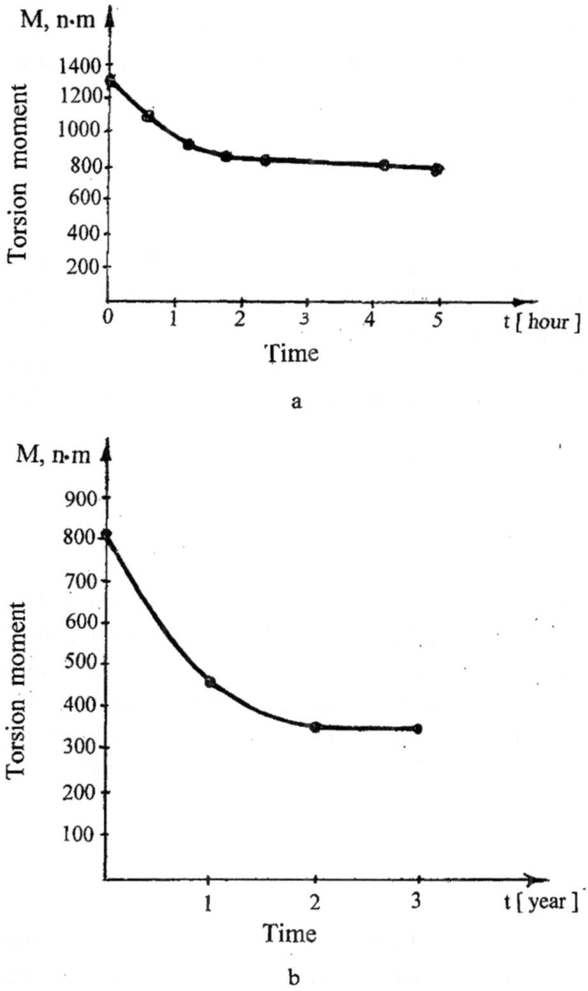

Fig. 22.13 4×4 convex lens array.

the lens shape. The lens is machined by a rotational tool around the outside of the lens. As a result, the cutting speed can be kept quite high, even at the center of the lens.

22.3.4.2
Micro Noh Mask

A Noh-mask shape was machined that is 1 mm in diameter and 30 μm in height. The Noh mask's shape data was created by digitizing the contours of a souvenir Noh mask. The tool used was a rotational pseudo ball-end-mill diamond with radius of 30 μm. The rotational axis of the tool tip was slightly offset from the rotational center to prevent the cutting speed from falling to zero. The work material was 18 K gold.

22.3.4.3
Micro Buddhist Model

Now, we are examining more complex machining shapes like the buddhist image shown here. We were able to machine this image even with a diameter of as little as 3 mm. The material was wax. In this example, the CAD-designed buddhist shape was machined by using the pseudo ball-end-mill described above. In the future, we hope to be able to machine even smaller pieces.

22.3.4.4
Micro Needle with High-Aspect Ratio

The next sample shows a needle that is 25 μm square and 1 mm high. The top of the square pillar has a sharpened angle of 15°. The rotational tool moves around and machines the needle shape. A surface is never machined more than once. As the machine vibration is very low and the cutting resistance very small, it is possible to perform machining without breaking the pillar, despite the shape being so long. The workpiece material was brass.

22.3.4.5
Micro-Injection Molding

In general, mechanical machining methods are not suitable for volume production, especially when compared with semiconductor processing. The latest trend is for precise dies fabricated by an ultra-precision machine to be used for plastic injection molding. Thus, ultra-precision micro-components will become cheaper and more widely used. We have an example of injection molding that uses a metal mold with micro V-projections with a width of 12.5 μm and a plate pitch of 25 μm. The next example is the injection molding using a metal mold with a V-groove that has a 1-μm pitch and a 140° V-angle, as introduced in 4-1. The plastic material is polyethylene. Although these injection-molding products seem to allow the injected plastic to flow into every corner of the mold, a micro *R* shape was observed at the bottom edge of the V-grooves. There is radius of about 0.1 μm. Now, we are studying how to reduce the micro *R* shape.

We have introduced a variety of micro-structures that were machined using high-accuracy mechanical machining. Micro-copy technologies will play a decisive role in bringing micro and ultra-precision devices into widespread use. Mechanical systems should make a comeback in the field of ultra-precision machining, and will be used both alone and in combination with semiconductor technology in the field of micro-precision fabrication.

22.4 Robot

22.4.1 Robotization

22.4.1.1 Automation and Robotization

The automation of oil refineries and chemical complexes is already as near-perfect as technology allows with equipment and instruments. Machining is automated by Computer Numerical Control (CNC), and assembly lines use conveyors and well-trained, skilled workers. More automation is expected in machining and assembly, but those operations performed by skilled workers require machines to be capable of human-like movement. Robots have been developed to realize these capabilities.

Early Motivation for Robotization. The drive toward robotization started thirty years ago with the following objectives, in order to change our manufacturing environment:

We expect robots to replace human workers in monotonous jobs in factories, such as loading/unloading workpieces onto machines or repetitious jobs in assembly. Most industrial countries, confronted with aging populations, declining birth rates and a tendency to pursue higher education, are concerned about labor shortages, especially in the so-called 3-D-dirty, dangerous, and demanding working environments. Robotization will relieve human workers of these jobs so that they can take on more rewarding jobs.

The population of workers with skills such as arc welding and casting has decreased in the above-mentioned labor market. Robots are expected to replace skilled workers in these jobs, thereby compensating for the shortage of skilled labor.

More flexible robots enable systems to produce many varieties of products in small lots, by changing the programming rather than the hardware. These production systems can be reused when product models are changed because robots, being generic equipment, are reusable.

22.4.1.3
Further Objectives of Robotization

Global competition has permeated all markets and has forced all manufacturers to reduce costs severely if only to remain competitive. Labor costs are one of the major factors affecting cost reduction. Many factories have been transplanted to areas where inexpensive labor is available, moving further and further away from the markets in pursuit of cheaper labor. Products have become complex, requiring more careful supervision at production. Customers also have become more alert to purchasing products that offer reliability. Robotization can provide solutions to these challenges.

- Robotization provides nearly full automation, thereby minimizing labor costs. Wherever they are, robots require the same amount of electric power around the world. They can achieve cost reduction wherever they are adopted.
- Robotization enables very precise manufacturing, which helps increase the quality of products. High efficiency in production often forces human workers to endure stress. Robotization can relieve them of such severe stress.
- Robotization enables round-the-clock automated operation without placing an unreasonable burden on human workers.
- Robotization increases the reliability of production systems because the robots themselves are so reliable.
- Robotization also minimizes the time needed to start up a production line because robots are very stable and well-prepared machines. Offline robot programming systems at offices as well as animated guidance tools on the factory floor are available to assist users in preparing the robots quickly.

22.4.1.4
Evolution of Robotization

The automotive industry has taken a very important role in acting as an incubator for robotics since their inception. Spot-welding robots for body shop lines are now very sophisticated. Robotization is now spreading more from the likes of spot welding, arc welding, and painting to areas where human skills are required, which calls for robots with more capabilities.

The key to providing robots with skills lies in creating machine intelligence by sensor-based control. Human beings have five senses, namely, sight, hearing, touch, taste, and smell. Out of these five senses, sight and touch are the most essential to performing jobs. Intelligent robots are equipped with vision sensors and force sensors. Vision sensors need to recognize objects not only in two but also in three dimensions, in order to allow robots to find objects and locate them, even if they have not been positioned precisely in advance. The vision sensors can eliminate peripheral equipment such as workpiece feeders. They can also relieve operators of the burden of having to place workpieces on the equipment. Force sensors enable robots to assemble parts without jamming them.

Many robots can be linked by a network to work together in a cooperative way to realize more capabilities and safe operation by avoiding collisions. An inexpensive and capable means of communication is now available for robots to communicate with each other in real-time, in order to synchronize their motion or measure the distances between them.

When a robot is combined with new processing technologies such as YAG lasers, the robotization achieves a synergy in the new manufacturing process. An optical fiber cable is mounted on the arm of the robot to transmit the YAG laser beam. The robot can precisely track a trajectory to enable cutting and welding. The YAG laser beam is focused onto a very localized point so that it does not cause the heat distortion seen in other processes such as arc or plasma welding. It also produces very smooth surfaces and eliminates the need for additional processes such as deburring after cutting, or grinding after welding.

One of the goals of robotization is having robots manufacture more robots, which is a very interesting way of reproducing machines. Figure 22.14 shows an example of "Robots manufacturing robots" with sensor-based intelligent robots. The system employs 3-D vision sensors to locate parts, while the robot hands use force sensors to assemble the parts. The robotization of manufacturing processes provides substantial flexibility and relieves human beings of dirty, dangerous, or demanding jobs, thereby allowing manufacturers to keep the production physically close to the markets rather

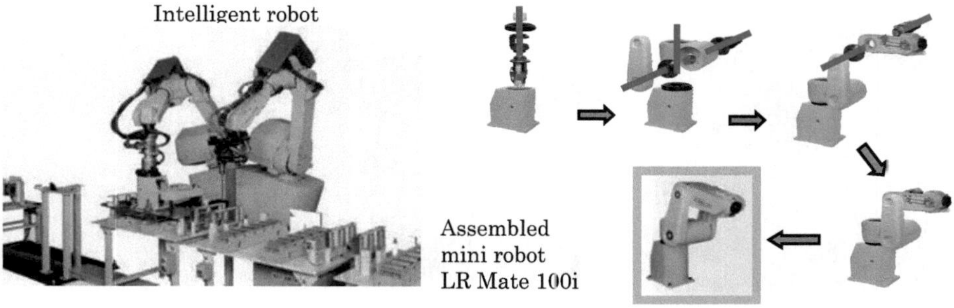

Fig. 22.14 Automatic assembly of mini-robots.

than transplanting their factories elsewhere in pursuit of inexpensive labor. Robotization is also indispensable for improving the reliability of products. Further research and development in robotics is called for to expand the potential of new products and new manufacturing processes.

22.4.2
Intelligent Robots

22.4.2.1
Objectives of Robotics Intelligence

Today, industrial robots are being used in many manufacturing processes, such as spot welding, arc welding and materials handling. However, there are some issues that remain unsolved:

1. Fewer peripheral devices: Current industrial robots require that the parts be fed to a precisely determined position. So, today's automated manufacturing systems require many dedicated and expensive peripheral devices.
2. Performing complicated assembly tasks: Assembly tasks often require the expertise and dexterity of a skilled human worker, such as fittings with very small tolerances, assembling gears, and so forth. These tasks are difficult for robots.
3. Reduction of time and effort needed for teaching: Currently, many man-hours are required for teaching robots.

Recently, intelligent robots have been drawing considerable attention as an effective means of solving these issues. Sensors are one of the key technologies for intelligent robots.

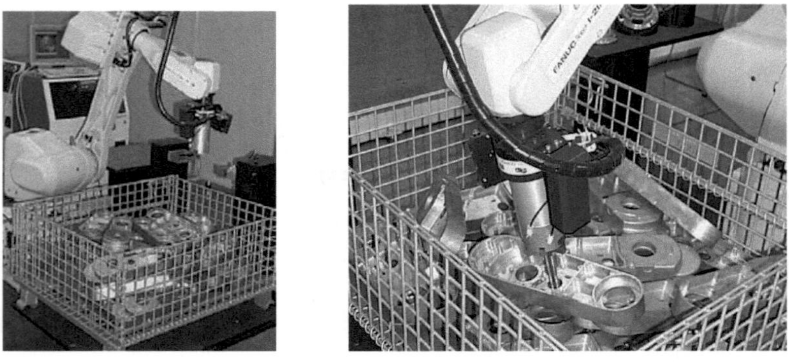

Fig. 22.15 Using a 3-D vision sensor to pick up parts from a random pile.

22.4.2.2
Using Vision-Sensors to Reduce the Number of Peripheral Devices

Three-dimensional vision sensors are used to locate a workpiece even in a roughly positioned pile. This ability means that there is no need for a peripheral device to hold the workpieces in precisely determined positions prior to the robots handling them. Figure 22.15 illustrates an intelligent robot with a three-dimensional vision sensor picking up a workpiece from a randomly filled box.

22.4.2.3
Using Force Sensors for Complicated Assembly

Intelligent robots equipped with force sensors can perform complicated assembly jobs such as fitting with a small tolerance, assembling gears, or assembling the clutch units of automatic transmissions. When fitting parts with a tolerance of less than 10 μm, the robot can fit them by sensing forces without jamming them. Figure 22.16 shows examples of robot assembly that rely on the use of force sensors.

Easy Teaching Assisted by Sensors. Figure 22.16 shows an example where a vision sensor and an offline programming tool are combined to facilitate the teaching of arc welding. A weld path can be specified on the image captured by the vision sensor. By tracking the weld path with the vision sensor, a robot program is generated automatically. This function saves an operator from teaching the robot how to perform the weld, even if they are an expert or not in teaching.

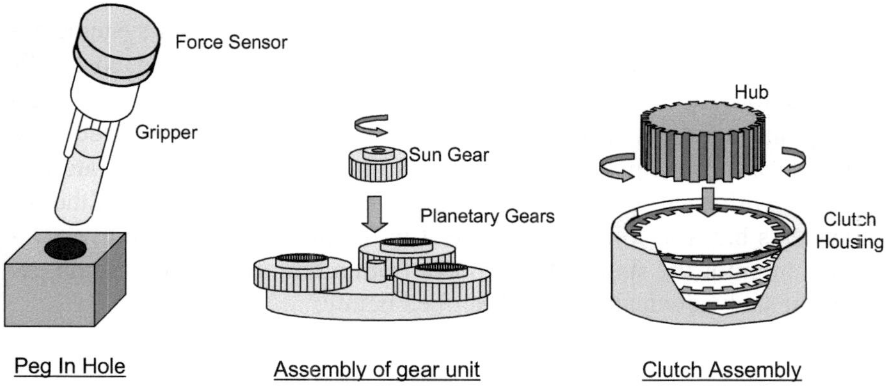

Fig. 22.16 Assembly with force sensor.

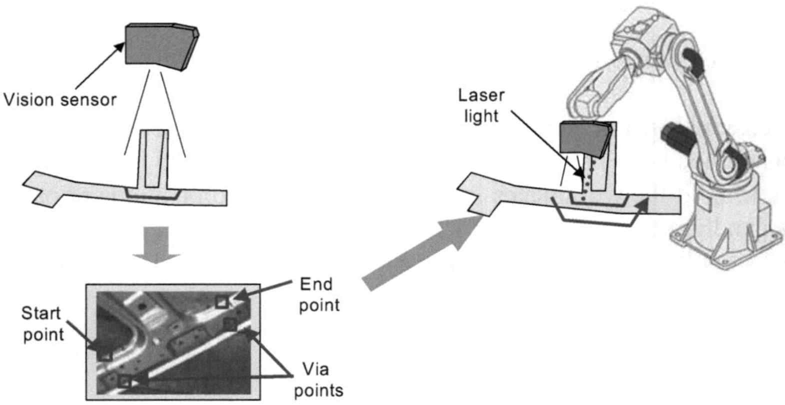

Fig. 22.17 Sensor-teaching function.

22.4.2.5
More Roles for Intelligent Robots Expected

Thanks to rapid advances in sensor technology, intelligent robots with sensors can perform complicated tasks requiring skills with relatively few peripheral devices and minimal teaching. Intelligent robots are expected to play an increasingly important role in the 21st century's manufacturing systems.

22.4.3
Networking

22.4.3.1
Networking in Robotics

Networking is becoming a common feature of industrial robot systems because of recent improvements in networking technology and cost efficiency. Before the adoption of networking, each robot was controlled individually and it did not know what other robots in the same workplace were doing. Networking can connect all robots and let them communicate and cooperate with each other. Networking can be extended not only to the robot systems but also to other equipment in the plants. Furthermore, operators can monitor the status of the robots from a remote location. Some examples of such new features are described below:

22.4.3.2
Network-Based Robot Controller

Each single robot is subject to limitations on the payload it can carry, as well as on its other capabilities. But, using networking to coordinate multiple robots could overcome such limitations. Such a cost-effective "flexible coordinated system" could replace existing expensive dedicated machines. The robot controller for such a "flexible cooperative operation" should be designed to offer a sufficiently high networking performance. The following figures (22.18–22.22) show a robot controller that was designed to achieve a sufficiently high networking performance. A dedicated processor is required for communication so that the processing related to networking never disturbs the robot motion.

22.4.3.3
Network-Based Coordinated Motions of Multiple Robots

New applications such as the handling of heavy materials or large parts – too large for a single robot – would be feasible by coordinating the motion of multiple robots. Coordinated motion requires close and frequent communication and good response between the robots. The latest robot controller offers a sufficiently high level of performance to enable this. With this

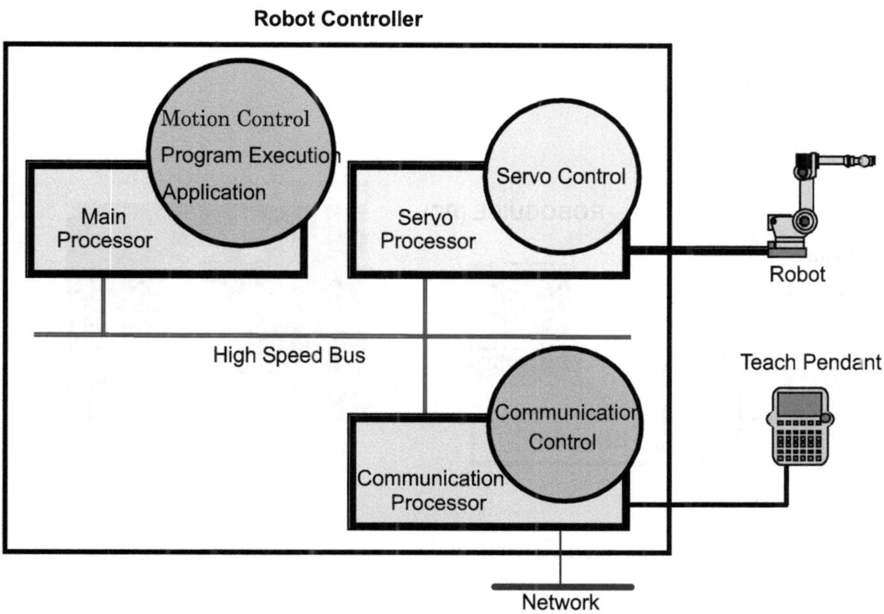

Fig. 22.18 Structure of network-based controller.

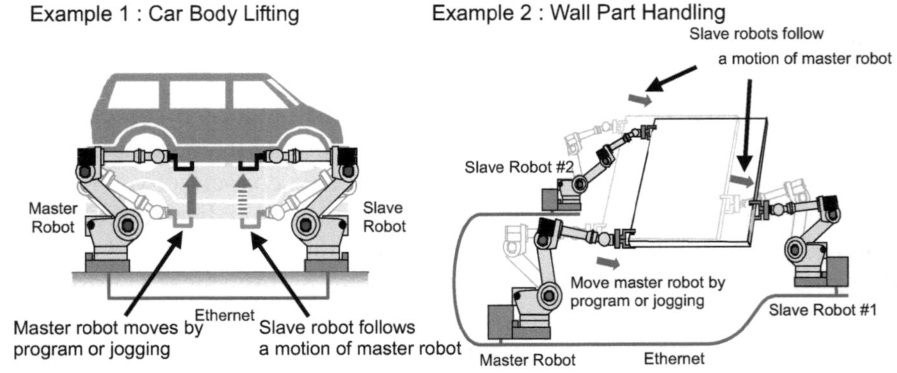

Fig. 22.19 Coordinated motion of multiple networked robots.

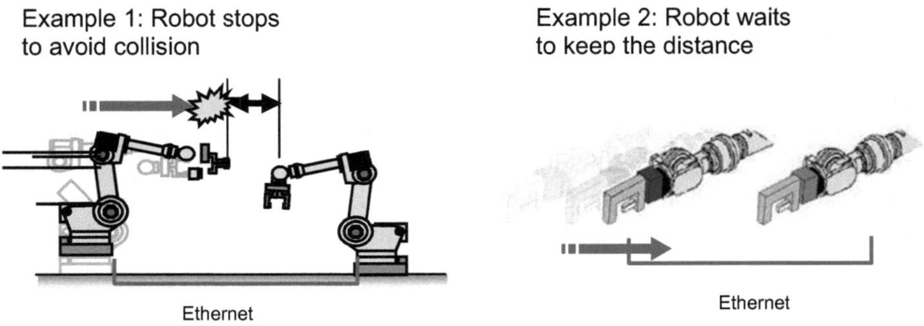

Fig. 22.20 Network-based anti-collision function.

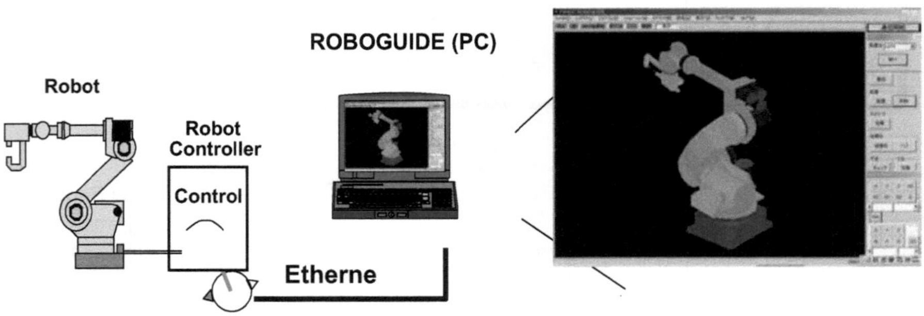

Fig. 22.21 Graphical human interface using PC.

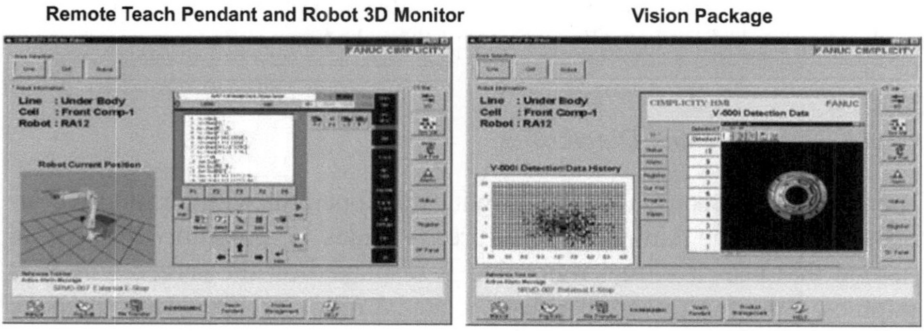

Fig. 22.22 Monitoring robot and vision data using CIMPLICITY.

system, only the master robot runs a program, and the other robots communicate with the master robot in real time and follow its motion to maintain a predetermined coordinated relation.

22.4.3.4 *Network-Based Anti-Collision Function*

Highly dense robot installations are very space efficient and make production lines shorter. To date, such installations have used a complicated interlock system to prevent collisions between the robots. With the new Ethernet network-based anti-collision function, the robots exchange information with each other on the shape and location of their arms and tools. Each robot can decelerate and stop before a collision not only in programmed operation but also in manual jog operation. Robots can automatically pause and resume their motion according to how far they are from a target.

22.4.3.5 *Better Human Interface Using Animated Guidance Tool*

Networking also improves communication between the operator and the robots. Previous robot controllers were specially designed to control robots with high performance and high reliability, but needed improvement in the interface between the operator and the robots. Networking allows us to connect a controller and PC tightly, so that a PC can be used as part of the robot controller. A better human interface can be provided with better graphics through the PC. And, because most of the processing is done by the PC, this does not affect robot motion performance or reliability. An example is connecting a robot controller to a robot arm or a PC. Graphic displays that animate the robot motion are realized using signals from the robot controller. This allows human operators to confirm the robot motion on the display before actually activating the robot arm.

22.4.3.6
Remote Operation and Diagnostics

The graphical screen on the PC provides the operator with an intuitive interface. And, the simulation feature provides a program verification method that uses a virtual robot that is displayed on the screen with 3-D graphics. You don't need to move the actual robot, and it is easy to check the parameters. The robot controller can be connected to the network of the plant, and the operator can monitor the online status of the robot through the network. The following shows an example of monitoring robot status using CIMPLICITY, a unique software product marketed for this purpose.

The network gets current information on the robot motion, which can be displayed on the screen using 3-D graphics. Furthermore, the robot can be operated through the network. In other words, remote setup and diagnosis are possible. If the robot is equipped with vision sensors, the captured vision data can be transferred to CIMPLICITY and all the data can be stored. This data can be used for analysis, for example, for analyzing all the data for a few weeks using statistical methods. This would provide a means of determining the reason for any vision-detection failures, thus improving the system.

22.4.3.7
Network Extending Robotics Capabilities

As we described above, networking would help overcome the existing limitations of the industrial robot. And, new applications could be explored based on this breakthrough. Networking would be extended everywhere within a manufacturing plant. This would make a plant a highly efficient system with better communications and complementary features.

22.4.4
YAG Laser Robot

22.4.4.1
YAG Laser Robot as a New Technology

Processing technologies that use lasers have already become commonplace. Especially, low-power Nd:YAG laser processing is used for marking, trimming, and spot welding electric/electronic parts, while CO_2 laser processing is used for sheet-metal cutting.

Recently, the number of applications of Nd:YAG laser systems (Fig. 22.23) has been rapidly increasing in machining-processing fields (mainly in the automobile and electric/electronic industries) because higher power Nd:YAG laser oscillators have been developed. Fiber optics can be used as the laser-beam transmission technology for Nd:YAG lasers, making it easier to equip a robot with YAG laser tools. YAG laser robots hold the promise of providing new robotics laser-machining processes. YAG laser robot systems are characterized by their compactness, easy maintenance, and wide range of motion.

22.4.4.2
Cutting Applications

In the laser-cutting field, laser robots are applied to flexible 3-D cutting with a high cutting quality that needs no additional processing after cutting (no dross and no burring). Usage examples include trims for the sheet-metal components and option holes in automobile assembly lines. Recently, 3-D flexible cutting by YAG laser robots was being applied to hy-

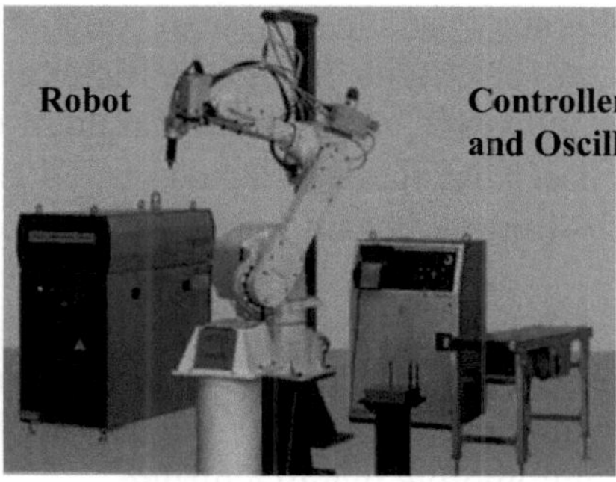

Fig. 22.23 YAG laser robot system.

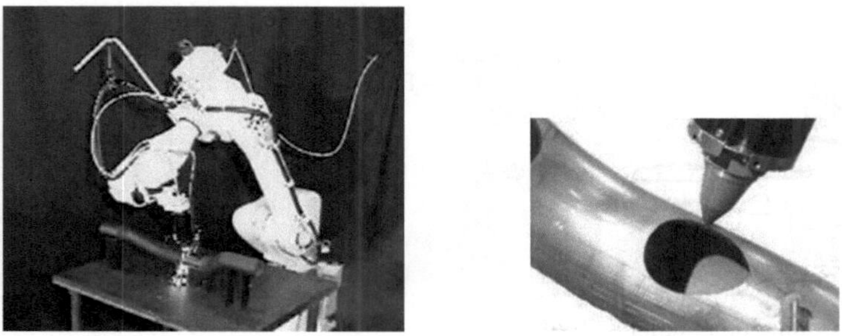

Fig. 22.24 3-D pipe (hydro-formed tube) cutting by laser robot.

dro-formed workpieces such as automobile frames. Hydro-formed workpieces cannot be cut prior to forming, and press cutting cannot be applied to 3-D workpieces. YAG laser cutting is ideal for such applications.

22.4.4.3
Welding Application

YAG laser processing is also used for high-speed welding without heat distortion (Fig. 22.25). In automotive body shops, high-speed YAG laser welding is already being used because it increases the welding quality, as well as the productivity and flexibility of the design.

The following example shows the application of YAG laser welding to the roof of an automobile. YAG laser welding increases the stiffness of the body and provides car body designers with more flexibility.

22.4.4.4
Key New Technologies for YAG Laser Robots

To date, a lamp-pumping method that uses discharge-arc lamp tubes has been the mainstream technology used in almost all Nd:YAG laser oscillators.

The energy conversion (electricity to light) efficiency of a laser-diode (LD)-pumping oscillator, however, is 10–20%, making it much higher than that of a lamp-pumping laser oscillator (2–3%). The adoption of LD-pumping laser oscillators will improve beam quality and the cost-performance ratio.

A large cutting range and the welding of 3-D shapes at high speed and with high quality can be realized by using a YAG laser robot with the LD Laser oscillator. In the near future, LD YAG welding technology will take over for spot welding in automotive manufacturing.

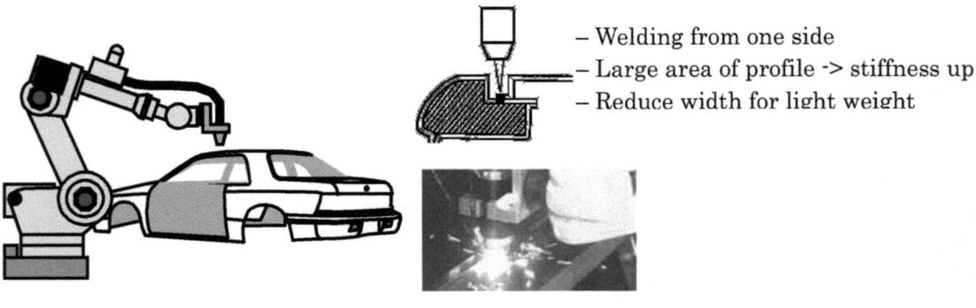

Fig. 22.25 Laser welding of automobile roof.

Fig. 22.26 Construction of LD-pumping laser.

22.4.5 Laser Processing and Laser Robots

22.4.5.1 What is a Laser Robot?

The arm of a laser robot carries a laser beam delivery system that emits a high-power laser beam that is used in material processing such as cutting, welding, surface modification, drilling, etc. This type of robot is often used in the automotive industry.

In laser material processing, the laser beam generated by a laser oscillator is first introduced through a beam-delivery system to the processing spot. It is then focused by a lens or a reflector onto a small spot, thus giving a high-power density. Finally, the beam heats the workpiece. This is a kind of thermal processing.

22.4.5.2 The CO_2 Laser – The Pioneer Laser

Scientists and engineers were aiming to realize this kind of application from the time the first ruby laser was developed in 1960. However, laser applications did not become popular until after laser oscillators, satisfying the following conditions, were developed.

First, a laser producing an output power that is high enough to be used for thermal processing is required. While the first ruby laser was used for drilling into diamond, people had to wait until lasers with greater capabilities were developed before they could be used in more everyday applications. The CO_2 laser is by far the highest power laser developed to date. Because of this, the CO_2 laser was expected to bring about a manufacturing revolution such that the attention of the people centered on this laser for a long time. Together with its high power capability, this laser has a high efficiency of 10% as the overall efficiency of conversion from electrical input to laser output. For these reasons, the CO_2 laser is today the most-often used in material-processing applications. However, YAG lasers, which are excited by the output beam from an LD are expected to surpass CO_2 lasers both in terms of output power and efficiency. Then, there will be another revolution in the field of laser-material processing.

Secondly, the wavelength of the laser to be used should be located in the interactive region between the laser beam and the work. In the thermal processing of metals, a beam of all wavelengths between ultraviolet and infrared can be absorbed easily, so that we don't have to pay attention to the selection of the wavelength. Both the CO_2 and YAG lasers can be used at will.

Thirdly, the technology of the laser to be used should be sufficiently mature. For lasers to establish their position as a useful tool, their reliability, cost, operational simplicity, and maturity of peripheral technologies should all be without question.

22.4.5.3
Differences from Conventional Technologies

The industrial advantages of laser processing include the novelty of the technology, the high processing quality, high productivity, high compatibility with automation, environmental improvement, high economy, safety, flexible processing, etc. The industrial disadvantages, on the other hand, include lack of expertise, questions of reliability, high initial investment, incompatibility with established manufacturing lines, etc. Which of these sides is more powerful will depend on the situation of the person weighing up the pros and cons.

22.4.5.4
Evolution from Flat Machine to Laser Robot?

Laser cutting found its first applications in the sheet-metal industry. Prior to this, simple pattern cutting had been done using a stamping machine, while complicated contour cutting had been done by laser. A hybrid machine comprising of the stamping part and laser part was used, which allowed a single machine to perform both kinds of cutting. Although the number of flat-metal laser-cutting machines surpasses that of three-dimensional machines, the latter is the better application of a laser. In the automotive industry, laser robots are being used increasingly, because they are ideal for the three-dimensional cutting of hydro-formed components.

22.4.5.5
Fiber Network Factory for Automotive Manufacturing

When YAG lasers come to be used more often than CO_2 lasers, the configuration of machines will change dramatically as the beam can be transmitted through an optical fiber. This helps to simplify the structure of a laser robot. The use of optical fiber for transmitting the beam means that the laser oscillator can be positioned well away from the processing spot. A network of optical fiber cables can be installed throughout the factory in the same way as electric cables. The laser oscillators do not have to be used in one-to-one correspondence to each of the processing spots. The laser light can be delivered in a time-sharing or beam-splitting manner to each of the processing spots. A YAG laser robot, which would have to manipulate only the processing nozzle attached to the terminal of the optical fiber, could be simplified in the way that it is controlled and, therefore, in its structure compared to a CO_2 laser-processing machine.

22.4.5.6
LD YAG as a Key Technology

Up until now, CO_2 lasers have been used most commonly in material-processing applications. The reason is that this type of laser dominates all others in terms of its output power, beam quality, energy-conversion efficiency, equipment size, etc. The wavelength of its beam is 10.6 μm, which is absorbed strongly by many metal and non-metal materials.

On the other hand, the YAG laser is a "diamond in the rough." All its properties, except for the fact that its beam can be transmitted through optical fiber, were until recently inferior to the CO_2 laser. Recently, however, the replacement of the flash lamps with LDs has led to a change in this situation.

It is now reasonable to expect that the conversion efficiency of the YAG laser will be increased by an order of magnitude. The efficiency, which is only one third that of the CO_2 laser, will become higher by a factor of two to three. An increase in efficiency will lead to a greater output power. The hardware size will decrease dramatically. The wavelength of 1.06 μm, which is one tenth that of the CO_2 laser, allows the laser to produce sharper cuts.

The CO_2 laser is a vacuum-tube device. Therefore, it requires mechanical components such as turbo blowers. The LD-pumped YAG laser, on the other hand, is an all-solid-state device, with no mechanical components. As electronics made huge progress after the transition from vacuum tube to transistor technology, the LD-pumped YAG laser will invite similar progress.

22.4.5.7 Lasers Change the Manufacturing World

Laser processing is a non-contact and inertia-free type of processing. The movement of the beam is free compared to a machine that is a large "lump of metal." In contrast to ordinary machines, which have only a limited amount of movement, a laser machine can be thought of as being "flexible." Any kind of processing can be performed by writing the appropriate software. Take a press machine as an example. A die needs to be prepared and the processing will be done using this die. In the case of laser processing, on the other hand, different kinds of products can be manufactured simply by switching the programming. Custom products can thus be manufactured at lower cost. When such products begin to hit the market, it is likely to have a huge impact on people's thinking and society as a whole.

22.5 Injection-Molding Machines

22.5.1 AC Servo Motor Injection-Molding Machine

Injection molding is one of the most productive methods of manufacturing. These days, not only plastic materials but also mixtures of metals or other solids and fluid can be injection molded, which widens the applicable field of injection-molding technology. In the early days of injection molding, the technology was usually applied to housewares and consumable products. These days, injection molding is also used for highly sophisticated engineering components, and the important parts of industrial products. Newly developed engineering plastics and materials allowed this to happen. But, they also forced changes in the injection-molding machines themselves.

The biggest differences in the new generation of injection-molding machines came from the difference in their control capability. Tighter tolerances, and very often, small, molded parts require precise and repeatable injection-process controls, demanding that pressure and injection speed are accurately controlled.

New materials are becoming increasingly heat-sensitive, such that it is necessary to accurately regulate each heating and cooling area. On the other hand, as the size of production runs becomes smaller, frequent mold-changing is necessary. So, machines must offer flexibility, smart mold changing, and easy startup.

The following describes the new technologies that are incorporated into the new generation of injection-molding machines.

The new trend started toward the end of the 1990s. Instead of using hydraulic pressure, increasingly injection-molding machine manufacturers chose to adopt AC servo motors. Not only small-size injection-molding machines with 500 KN of clamping force, or the 150-KN machines, but also 10,000KN and even bigger machines now use servo motors instead of hydraulics. More than 50% of the injection-molding machines sold in Japan during 2000 were driven by AC servo motors (Fig. 22.27).

The first injection-molding machine to be driven by a servo motor appeared at K'83 in Dusseldorf, but was no more than a prototype at that time. In 1984, FANUC launched the AUTOSHOT, our first realistic fully electric injection-molding machine, as the Japan's biggest leading manufacturer of AC servo motors and numerical controllers. It took another decade before injection-molding machines driven by AC servo motors came to be recognized as being on a par with hydraulic machines.

Injection-molding machines that are driven by AC servo motors have good characteristics in terms of energy efficiency, consuming only 1/3 of hydraulic-injection machines. They also offer accurate velocity and position control and stable operation, require less frequent maintenance, and are easier to work on. Nowadays, derivative technology, such as metal-press machines and die-casting machines that are entirely driven by AC servo motors, are also becoming popular.

The appearance of these machines did much to widen the field of plastic-mold applications. Precise and non-spherical plastic lenses can now be manufactured at very reasonable costs. Ultra-thin wall plastic cases allow mobile computers to be thin and light. As the years pass, we expect to see

Fig. 22.27 AC servo motor injection-molding machine.

more and more hydraulic-molding machines being replaced by those that use AC servo motors, much like hydraulic robots were replaced by those using AC servo motors.

22.5.2 Pressure-Profile Trace Control

Pressure-profile trace control, one of the special features of FANUC's 'ROBOSHOT' AC servo motor-driven injection-molding machine, is a typical example of the features that contribute to the capabilities of an AC servo motor-driven injection-molding machine.

Key to the success of injection molding is known as how the plastic material melts, flows, and solidifies in the mold. To analyze the solidification process, a pressure-volume-temperature (PVT) curve for each material is essential. It shows how injected plastic material solidifies in a die to produce a plastic part. The actual quality of each mold part depends heavily on the pressure curve during the injection process. Thus, when the injection pressure can be controlled in individual time periods, ensuring the optimum conditions shall result in a high level of quality.

The pressure-profile trace control is one of the AI features of the ROBOSHOT. With this method, the pressure profile during the injection period is directly commanded and closed-loop controlled. Stable injection molding can be obtained through this method, even if an external violation occurs, such as a change in the material characteristics or a change in the external temperature. The pressure-profile trace control is one of the technologies that made AC servo motor-driven injection-molding machines possible (Fig. 22.28).

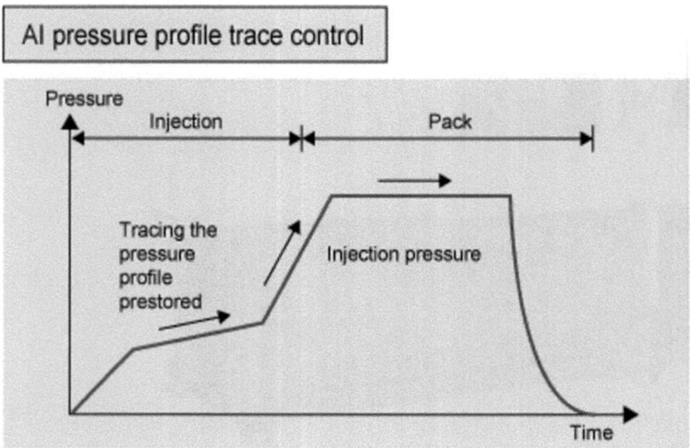

Fig. 22.28 Pressure-profile trace control.

22.5.3
Centrally Controlled Injection-Molding Factory

On the other hand, injection molding is a very efficient way of production but, at the same time, it is prone to certain problems, even in mass-production applications. Problems can be caused by the injection-molding machines themselves, material, mold, an operator, and so on. Upt-to-date injection-molding factories use a central monitoring system for enhancing productivity and quality. The following systems are already in operation:

- Central monitoring of machine operation.
- Central quality control.
- Central production control.
- Remote service and maintenance.
- Molding-condition-adjusting expert.

Figure 22.29 shows the FANUC MOLD24*i* system, which is mainly capable of central monitoring, quality control, and production control with only a small extra cost. Injection-molding machines (ROBOSHOTs) are tied to a central personal computer through an Ethernet network, while each ROBOSHOT has a communication capability as standard. Round-the-clock remote service and support is provided through an Internet connection between FANUC's service center and each individual customer.

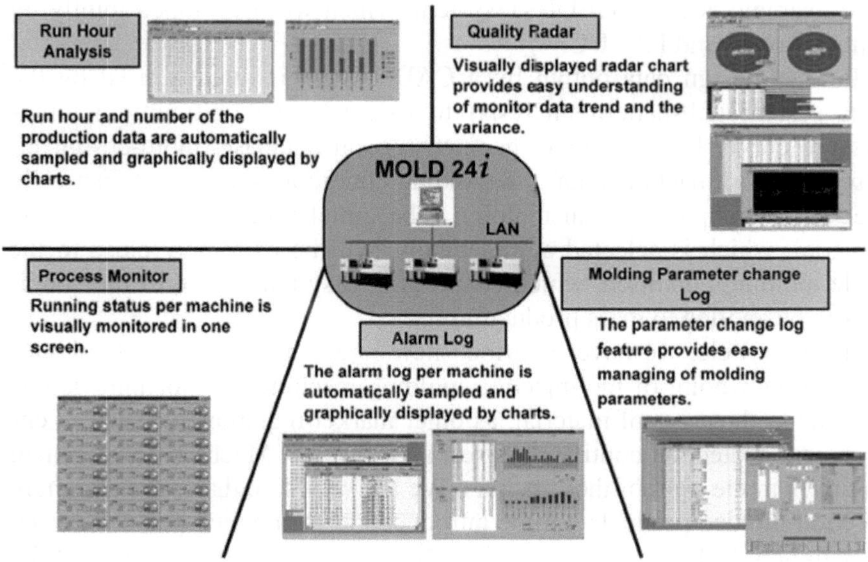

Fig. 22.29 FANUC monitoring and control system.

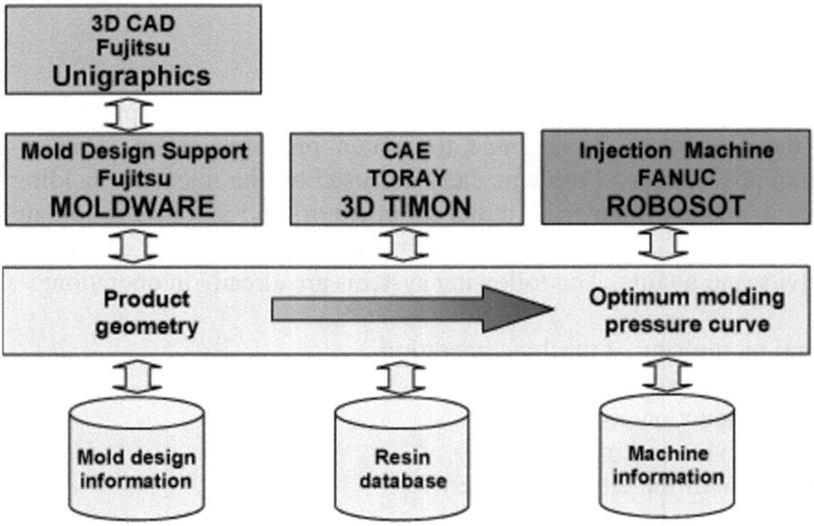

Fig. 22.30 MOLDEST system.

22.5.4 Integration of CAD, CAM , CAE and Injection Molding

The wave of digital engineering, which is at the core of IT application to manufacturing, is now reaching the field of plastic injection molding. One good example is the MOLDEST system, which was developed jointly by Fujitsu, Toray, and FANUC.

Product-design data output by a CAD system is directly used by the CAE system (fill-pattern analysis) to calculate the optimum molding-pressure curve. Using product geometry, resin database, mold-design information, and machine data, the molding conditions can be automatically selected to complete virtual molding. The initial setup conditions for the first trial, which is selected automatically, will be extremely close to the stable molding condition, which enables the easy realization of the conditions to be applied to mass production.

In this article, we have described new injection-molding technologies from the viewpoint of the injection-molding machines. Major innovations continue in the areas of material, as other market requirements such as environment protection continue to exert an influence. Machine development continues in the light of these factors. We are very sure that more and more injection-molded products will be introduced in the coming years. To keep up with new technologies, we will continue to answer the demands of the market.

22.6
Wire Electric-Discharge Machines

22.6.1
About Wire Electric-Discharge Machines

Wire electric-discharge machines use discharge energy to continuously melt metal. While feeding out an electrode wire (approx. 0.05–0.3 mm in diameter), the machine performs two-dimensional contouring in a manner similar to jigsaw cutting. These machines were invented in the U.S.S.R. in 1960. At that time, cutting was performed by manually moving a table backwards and forwards. During the 1970s, numerical control of the table movement was realized, substantially increasing the practicality and cost-effectiveness of the machines. Figure 22.31 shows a wire electric discharge machine.

Wire electric-discharge machines can handle any complex contour that can be drawn with a single stroke, and can easily achieve a cutting precision (0.01 mm or less) that is one degree of magnitude better than any gen-

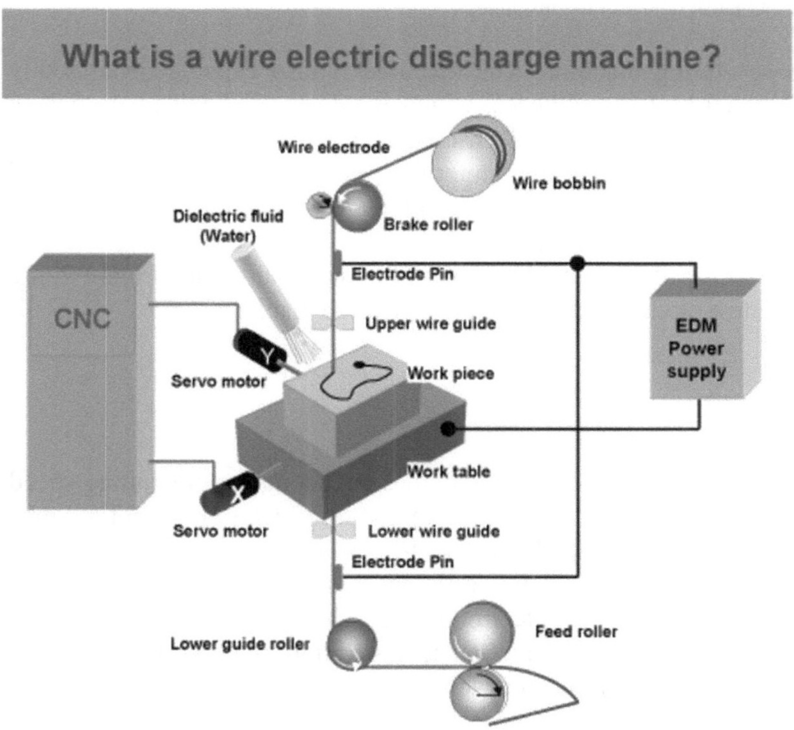

Fig. 22.31 Structure of a wire electric discharge machine.

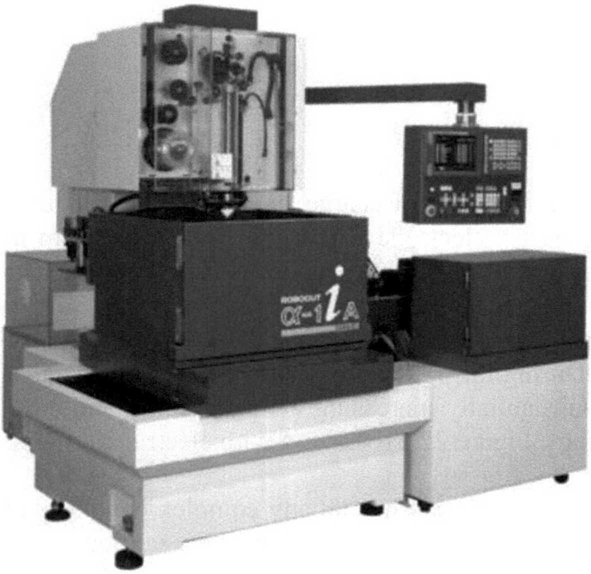

Fig. 22.32 The latest model of the machine.

eral-purpose cutting machine. The machine is also capable of smoothly cutting high-hardness workpieces, which would be difficult to do with a general cutting machine. Wire electric-discharge machines are mainly used for cutting metal dies (pressing, sintering, plastic molding, extrusion, and draw dies), electric-discharge machining electrodes, prototype and mass-production parts, gages, and edged tools. The market for this type of machine sees worldwide annual shipments of about 6,000 units. Figure 22.32 shows the latest model of wire electric-discharge machine.

22.6.2 High Speed

Currently, wire electric-discharge machines have a maximum cutting speed of approximately 300 mm^2/min (table feed-rate of 5 mm/min and a workpiece thickness of 60 mm). To speed up cutting, the maximum possible discharge energy must be supplied efficiently to the electrode gap (gap between the workpiece and the electrode wire) while avoiding electrode wire breakage. For this purpose, the pulse-generation method was recently changed from the traditional capacitor-based type to a transistor-based type, with which discharge pulses can be generated more abruptly and at a higher frequency.

Wire electric-discharge machines allow unattended, easy cutting with a high degree of precision. No manual finishing is required after cutting, because the machine performs non-contact cutting and therefore ensures that the workpieces remain free of burrs, which would otherwise be generated during cutting. For this reason, when a higher cutting speed is achieved by using improved discharge-pulse generators and electrode wires, it is likely that a growing number of wire electric-discharge machines will be used for the cutting of mass-production parts, probably expanding their market size.

22.6.3
High Precision

In metal die cutting, one of the major uses of the wire electric-discharge machines, a significant challenge is the improvement of the precision. The evaluation criteria for the cutting precision are classified into the following, as shown in Fig. 22.33: <1> The straightness precision resulting from the discharge device operation, <2> the profile precision resulting from the machine precision, and <3> the corner precision resulting from the machine control. A mating die and punch with a complex corner requires mechanical feeding in all directions, as shown in Fig. 22.33, and cutting must be performed with the highest degree of precision.

To improve the straightness precision, sludge (chips) in a cutting groove and the intensive discharge caused by wire vibration should be removed. To this end, a technology has been developed to detect the discharge status in real time and so control the discharge pulses, thus maintaining a constant discharge in the workpiece direction. The latest technology for profile precision improvement is fully closed control with linear scaling.

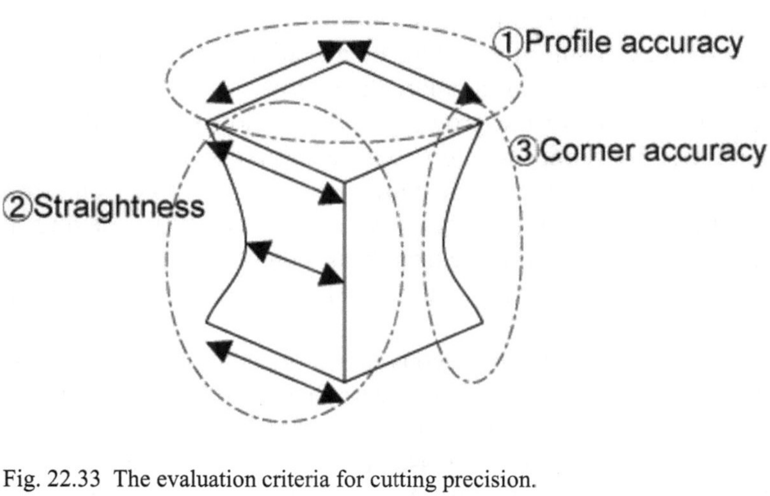

Fig. 22.33 The evaluation criteria for cutting precision.

Fig. 22.34 Result of machining.

During cutting, the electrode wire tends to be deflected by the flow of a dielectric fluid. If the cutting direction is changed while this deflection is present, a corner can be cut excessively. To avoid this, operators would have to draw on their experience to adjust the intensity of the discharge and change the cutting paths, based on corner profiles or angles. Currently, the control unit can read programs with paths specified in advance, to automatically apply optimum control at each corner, so that even inexperienced operators can achieve the best corner precision.

These technologies enable the cutting of a mating die and punch, as shown in Fig. 22.34, with a gap of no more than 2 μm.

In recent years, to satisfy the growing need for downsized industrial products, as typified by semiconductors, wire electric-discharge machines have been developed to support finer cutting with higher precision. Accordingly, smaller-gage electrode wires have been used. Wires having a minimum diameter of 0.05 mm are currently in practical use. Studies are now being made on high-precision cutting with electrode wires having a smaller diameter (0.03–0.02 mm).

Studies are also being made on gaseous-discharge cutting, which seems set to be a future cutting technology. The wire electric-discharge machine discharges electricity through a dielectric fluid that cools the workpiece and carries away the sludge (chips). Gaseous discharge will use a gas instead of a dielectric fluid (water or oil). Although the cutting speed may be lower, the discharge repulsion is small, thus reducing the wire vibration. Gaseous discharge is thus expected to be a means of providing very high-precision cutting.

22.6.4 Unattended Operation

The automatic wire-feed function is indispensable to the unattended operation of a wire electric-discharge machine. For a progressive die, like that shown in Fig. 22.35, the electrode wire must be disconnected after the cutting of the profile. Then, the wire must be moved to the cutting start-hole for the next profile, and reconnected. The automatic wire-feed function enables the machine to automatically perform wire disconnection and reconnection, without the intervention of an operator.

The reliability (connection rate) of the automatic wire-feed function is highly dependent on how the wire is disconnected. While a mechanical-disconnection method has been used to date, this causes variations in the shapes of the disconnected wire ends. As a replacement for this traditional method, a fusion method is becoming increasingly popular. With this method, current is fed through an electrode wire under tension until the wire is fused. With the wire-fusion method, the disconnected wire ends are burr-free and sharp, considerably increasing the reliability of the automatic wire feed. Therefore, unattended operation can be performed reliably and over a long period of time.

To prolong the unattended operation of a wire electric-discharge machine, another challenge involves automating the loading and unloading of

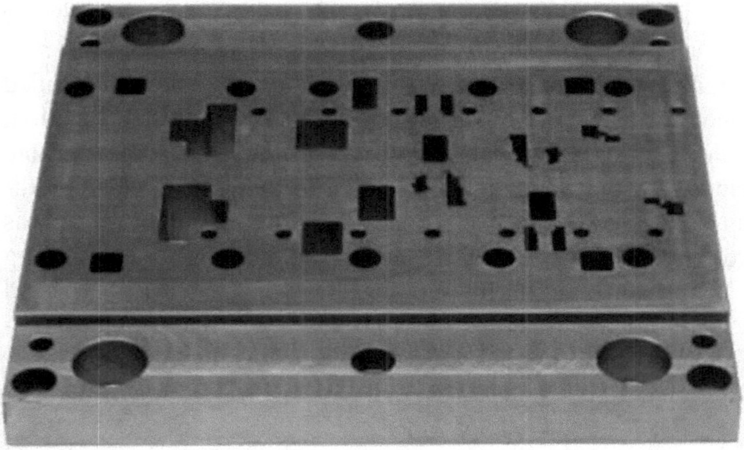

Fig. 22.35 Example of a part.

the workpieces. To automate this aspect, robot-based methods and pallet changer-based methods have both been suggested. In the future, as cutting speeds increase, it is likely that a growing number of wire electric-discharge machines will be used for the cutting of mass-production parts. Then, it is probable that automated lines will be constructed, each having multiple wire electric-discharge machines and a workpiece loading/unloading robot running from machine to machine.

22.6.5 Networking

In recent years, as personal computers have become highly sophisticated and commonplace, networks have come to be used more and more with wire electric-discharge machines. Major developments in networking include <1> a function for realizing high-speed NC-program transfer, and <2> a monitoring function for realizing long-term unattended operation.

Previously, NC programs were stored on media such as tape or Floppy Cassette and were transferred to the wire electric-discharge machines over an RS-232C interface. However, as cutting profiles have become more complicated, an increasing number of NC programs have been created with CAD/CAM, which demands an enormous amount of storage.

Many manufacturing sites have built LANs to transfer such huge programs to their machines. A wire electric-discharge machine with an Ethernet board installed can be used as part of a LAN. In this way, networking enables the high-speed transfer of a large number of CAD/CAM-based programs to the machines.

With the latest models of wire electric-discharge machines, advanced automation technologies allow continuous unattended operation for longer than 100 hours, and it is now common practice for a factory to be left running, unattended, over the weekend. To support this extended unattended operation, a monitoring function has been developed. Using this function, the operator can check the operating status of a machine from a remote personal computer, without having to go down to the shop floor. In addition, IT evolution has led to the instigation of a mailing function, as well as a function for automatically sending reports on the machine-operating status or any abnormalities to an Internet-capable cell phone. It is expected that, in the near future, manufacturers will provide users with troubleshooting, parts sales, and other services over a network.

22.7
State-of-the-Art FANUC Factories Incorporating FA and Robot Technologies

FANUC is a general manufacturer of FA and robot systems, including industrial controllers such as CNCs and PLCs, servo motors, robots, wire-cut EDMs, CNC drills and electric injection-molding machines. FANUC manufactures these products in state-of-the-art factories that themselves incorporate FA and robot technologies – providing a showcase for the actual application of FA and robot technologies. The section below introduces four of FANUC's factories, namely, the machining factory, servo motor factory, robot factory, and sheet-metal factory.

22.7.1
Machining Factory

The machining factory machines parts for robots, wire-cut EDMs, CNC drills, and electric injection-molding machines. When it started operation in 1980, the factory was Japan's first FMS factory to run unmanned at night.

Now the factory is equipped with state-of-the-art "FANUC CELL 60" systems that run unmanned for 72-h stretches through the weekends. The FANUC CELL 60 consists of 4–5 machining centers, a washing station, robots, and pallet stocker. An automatic supply of jig-mounted workpieces and fixtures to the machining centers enables the factory to operate unmanned at night and even through the weekends. At the robot station, FANUC robots perform deburring and washing of the workpieces. Figure 22.36 shows a FANUC CELL 60 in the factory.

22.7.2
Servo Motor Factory

The servo motor factory produces servo-motor and spindle-motor assemblies. The factory features a level of robotization that allows us to maximize its operating hours to include nights and holidays, in addition to normal working hours.

Every motor-assembling process, from parts retrieval, through assembly and inspection, to transportation, is automated. There are about 300 FANUC robots in the factory, with the ration of robots to operators being 7 to 1. Intelligent robots equipped with vision sensors perform jobs that have previously been difficult to automate. These robots have extremely powerful functions and are capable of tasks such as precisely taking parts from a pallet and inserting connectors, making full use of their vision sensors.

Fig. 22.36 FANUC CELL 60.

Fig. 22.37 Servo motor factory.

There are several motor-assembly lines, each of which produces a different range of motor sizes. The line for assembling small- to medium-size spindle motors is fully robotized from assembly to final inspection. On the line for assembling larger spindle motors, motors up to 150 kg are assembled with maximum efficiently.

The longer operating hours give a lower factory cost per hour, that lets us produce servo and spindle motors at highly competitive prices. Processes are robotized wherever possible, to produce motors of a consistent high level of quality while reducing lead-time. Figure 22.37. shows the servo-motor factory.

22.7.3 Robot Factory

The robot factory undertakes the complete production of FANUC robots from parts receiving and inspection, through assembly and testing, to shipment.

Fig. 22.38 Two-armed intelligent robot.

Fig. 22.39 Sheet-metal factory.

In the factory, FANUC's two-armed intelligent robot automatically assembles multi-purpose mini-robots with no human assistance. FANUC's two-armed intelligent robot is equipped with 3-dimensional vision sensors and 6-axis force sensors for recognizing randomly oriented positions and orientations, and to perform precision fitting and screw tightening with a degree of precision equal to a skilled worker. This is an innovative automated assembly system that realizes the goal of robots building other robots. Figure 22.38. shows a two-armed intelligent robot in the factory.

22.7.4 Sheet-Metal Factory

The sheet-metal factory is a state-of-the-art factory for producing cabinets for FANUC robot controllers.

Sheet metal is retrieved from the material stock shelves by a robot and cut into cabinet parts by a complex laser-punch machine. These parts are transferred to press brakes to be bent by robots. The bent pieces are delivered to the following robot stations that perform welding-assembly, powder painting, and final assembly. After these processes, the cabinets are completed.

Within the factory, the most appropriate FANUC robots are adapted for each process, allowing the entire factory to be run by only two operators. Figure 22.39 shows the sheet-metal factory.

Acknowledgments

Part of the work in Section 22.3 was performed by FANUC LTD and the laboratory of Prof. Takeuchi of the University of Electro-Communications. Part of the work was performed by FANUC LTD under the management of the Micromachine Center as the Industrial Science and Technology Frontier Program (ISTF) of the Ministry of Economy, Trade and Industry (METI), supported by the New Energy and Industrial Technology Development Organization (NEDO).

Literature

[1] Inaba H (1999) Newly developing control theory toward practical use. Measurement and Control 38 (1):4–9

[2] Inaba H (1998) Flexible automation technology with robotization and intelligence. Japan USA Symposium for Factory Automation, Ohtsu

Chapter 23
Perspectives of Innovative Technologies in Manufacturing

J. Gausemeier

23.1
Active and Intelligent Systems

The majority of mechanical engineering products today are characterized by a close interaction between classical mechanics and electronics, control engineering and software, expressed aptly by the term mechatronics. However, further possibilities are emerging that extend way beyond the scope of mechatronics – systems with inherent intelligence. Future systems in the area of mechanical engineering will comprise configurations of intelligent system elements, which we also refer to as solution elements since they enable the solution to a function. The performance of the overall system is characterized by the communication and cooperation between intelligent system elements. In terms of software, this involves distributed systems of interacting agents.

An agent is an autonomous, proactive, cooperative, and extremely adaptive function module. Autonomous implies an independent control system, which itself initiates actions (proactive). Agents are regarded as function modules, which work in cooperation or competition with one another. Adaptive identifies generic behavior at runtime, which may also, for example, include a learning capability. A function module is taken to be a heterogeneous subsystem with electronic, mechanical, and IT-related components.

This section introduces the basic structure of mechatronic products and the structure of adaptive and intelligent systems.

23.1.1
Mechatronic Systems

Modern mechanical engineering products comprise a basic system – i.e., a mechanical structure, sensors, actuators, and online information processing, which is generally performed by a microprocessor with software. Figure 23.1 shows the relevant basic structure of a mechatronic system. The status variables of the basic system are measured by sensors. The information is passed on to the information processing units and the necessary effects are determined. These different effects are then implemented directly in the basic system by actuators.

The control system depicted in Fig. 23.1 can act as the basis for creating iterative structures in terms of system theory, as clarified for example using a reflex camera. Thus, an actuator like the ultrasonic motor of the "autofocus" camera subsystem also represents a basic structure of a mechatronic system. It is helpful to depict the relationships between the components using flows so that we can examine the links symbolized by the arrows more closely. In essence, three types of flow can be distinguished: material flows, energy flows, and information flows (Pahl and Beitz 1997), whereby the information flows are frequently also referred to as signal flows.

The basic system of a mechatronic system comprises units linked via all three types of flow (Fig. 23.2). Energy and material flows are generally in the foreground in this context. The flows that link the basic system with the sensors and the actuators have the character of both energy and information flows. This is because energy "flows" both for measuring (sensors) and influencing (actuator), but also because information – control information from the actuators and measurement information from the sensors – is transmitted.

Both in relation to *sensors* as well as in relation to *actuators*, there is a tendency today to integrate sensors and actuators physically with other function units. Integratable function units include analog/digital converters

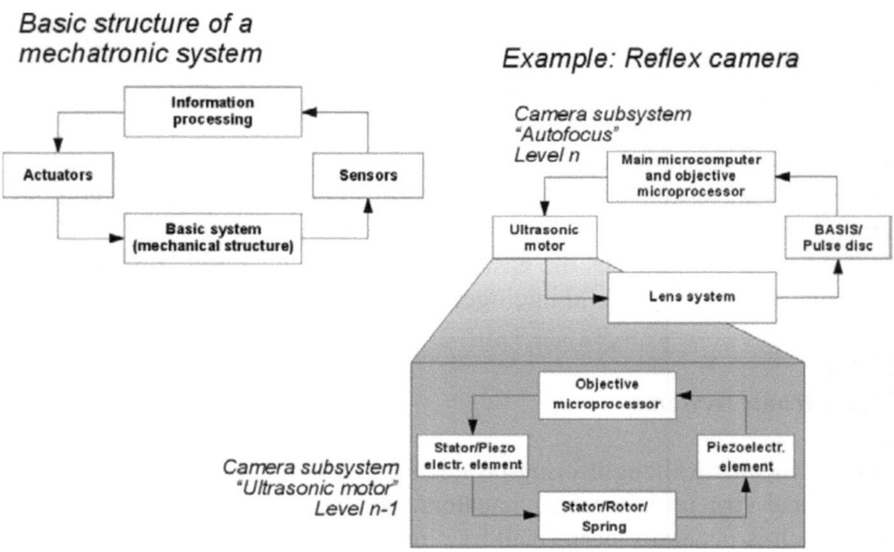

Fig. 23.1 Basic structure of a mechatronic system.

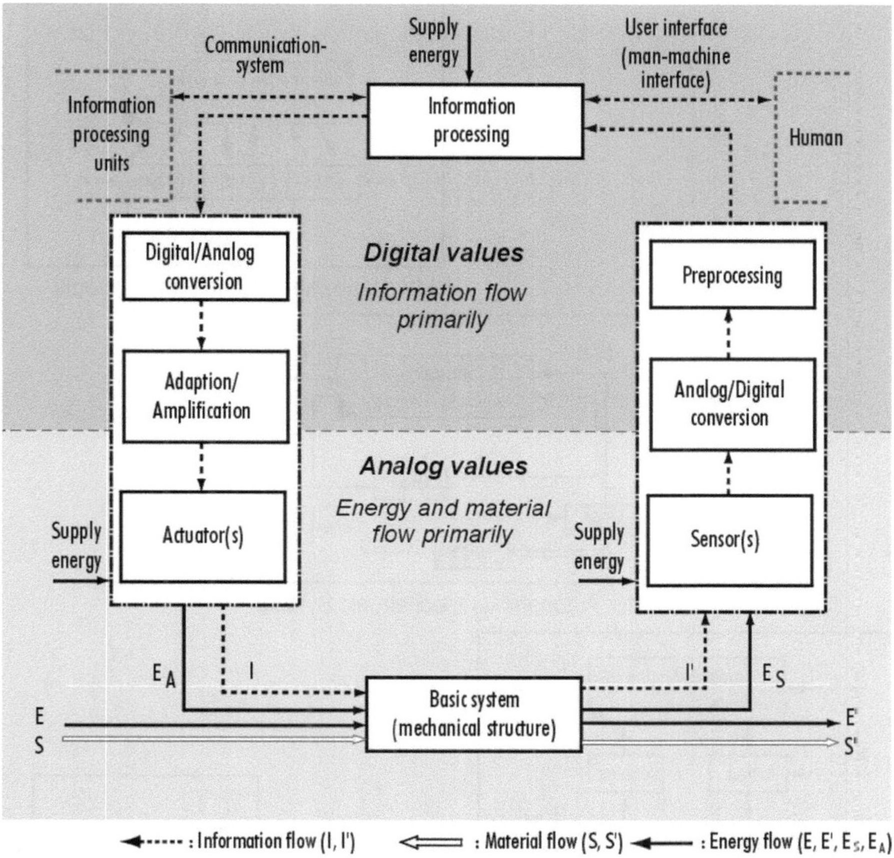

Fig. 23.2 Linking the basic system, sensors, information processing, and actuators via material, energy, and information flows as well as the basic structure of intelligent sensors and actuators.

and microprocessors for "intelligent" sensors as well as digital/analog converters and adapter circuits and booster circuits for "intelligent" actuators. "Intelligent" sensors measure physical analog variables such as pressure, temperature, or speed, digitize the measured values, and transmit the appropriately adapted signals to the information processing units. "Intelligent" actuators are controlled directly by digital information-processing signals. The signals are converted to analog variables, amplified, and are then used to create motion or energy. Sensors and actuators thus take over the conversion between energy and information flows with ease (cf. Fig. 23.2). The *information processing units* use the sensor information to achieve predefined goals and create control information for the actuators.

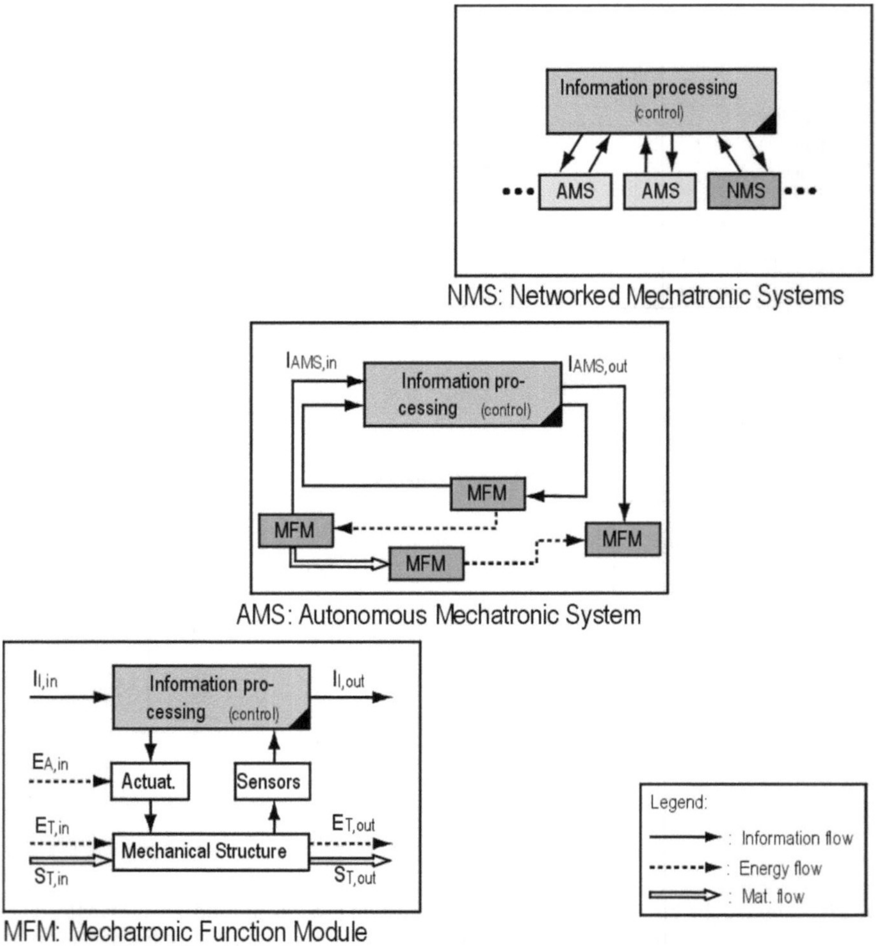

Fig. 23.3 Hierarchical structuring of mechatronic systems to master complexity.

In addition, the processing units are often interlinked via a communication system or communicate with the user via one or more user interfaces.

A further form of hierarchical structuring is proposed in accordance with Fig. 23.3 (Lückel 1992, 2000), in relation to mastering the extreme complexity of mechatronic systems. This solution is based on so-called *mechatronic function modules* (MFM), which consist of a supporting structure, sensors, actuators, and a local information-processing unit. *Autonomous mechatronic systems* (AMSs) are made up of MFMs linked on the basis of information technology and/or mechanically and include the respective sensors and the related information processing. The latter handles higher level tasks, such as monitoring with error diagnostics and decisions regarding maintenance, and generates defaults for local MFM information processing. AMSs form so-called *networked mechatronic systems* (NMS). NMSs are created simply by linking the participating AMSs via information processing. As with AMSs, higher level tasks are implemented in information processing by NMS. Using the example of motor vehicle technology, active MacPherson structural units would be MFMs, vehicles with mechatronic chasses would be AMSs and vehicle groups would be NMSs.

While the components of the basic system are interrelated primarily on the basis of energy and material flows, and thus also on the basis of analog, i.e., continuous value and time variables, the information processing units today operate almost exclusively on the basis of digital, i.e., discrete-value and discrete-time, variables. Intelligent sensors and actuators facilitate routing between the different physical areas.

23.1.2 Intelligent Systems

A prominent feature resulting from the integration of information technology in mechanical engineering products is self-optimization. The following explanations are based on a research project carried out at the University of Paderborn with the title "Self-Optimizing Systems in Mechanical Engineering" (heads of project: Prof. Gausemeier, Prof. Lückel).

The examination of self-optimizing systems is based on the four aspects of target system (e.g., target hierarchy or target vector), structure (i.e., topology of mechanical components, sensors, and actuators), behavior, and parameters (Fig. 23.4). Self-optimization is characterized accordingly by two features:

- The endogenous modification of the target system based on changing influences on the technical system;
- The targeted, autonomous adaptation of parameters, behavior, and structure.

The structure of self-optimizing systems is based on the structure of mechatronic systems, in that self-optimizing information processing (operator) is superimposed on the controlling mechatronic information processing (controller) (cf. Fig. 23.5).

The task of the operator is to extend the controlling system towards an intelligent controller behavior. Thereto the operator determines the necessary adaptations to changing working and environmental conditions, e.g., malfunctions, changing temperature, or humidity. These adaptations of the

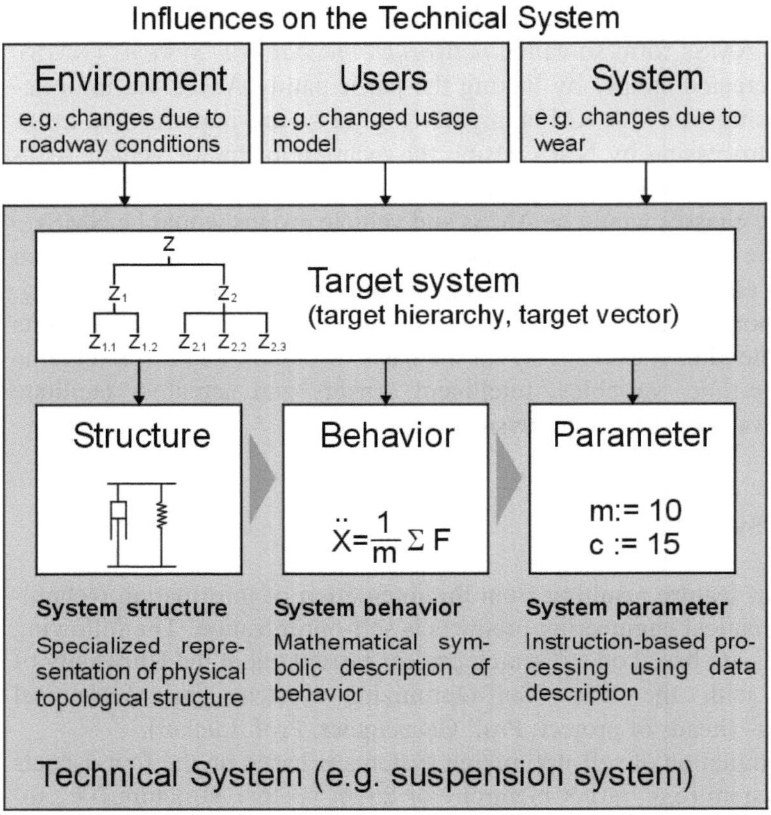

Fig. 23.4 Self-optimization enables empowered systems with inherent "intelligence," which are able to react autonomously and flexibly to changing environmental conditions.

controller behavior are transferred to the controller via events (Naumann 2000).

Self-optimizing information processing predefines the target system, the structure, the behavior, and the parameters for controlling information processing and requires two categories of knowledge for this purpose:

- Knowledge about possible target systems, structures, behavior, and parameters.
- Knowledge about selection or adaptation processes for parameters, behavior, and structure.

The following four aspects should be emphasized in particular in terms of the structuring of mechatronic systems in MFMs, AMSs, and distributed mechatronic systems (DMSs):

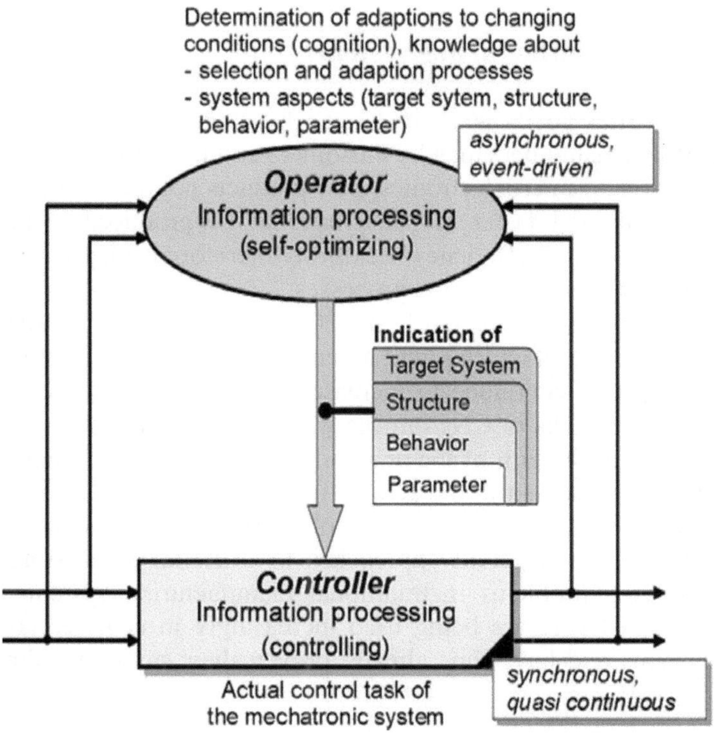

Fig. 23.5 The overlaying of the controlling information processing by the self-optimizing information processing leads to the creation of self-optimizing systems.

- *Self-optimization at all levels:* Self-optimizing information processing can take place at all levels of mechatronic systems (i.e., MFM, AMS, NMS). Based on this, self-optimization mechanisms can be implemented at every level and generalized sufficiently through structuring to allow the creation of reusable self-optimizing modules.
- *Proximity to actuators:* The mechatronic systems examined are structured hierarchically. The actuators are addressed at the lowest level, i.e., at the level of mechatronic function modules (MFM).
- *Response times:* From a current perspective, "fast and secure" self-optimization will be resolved through parameter switching, with changes also having to be made in a predictable way. In contrast, there will be a slower self-optimization, i.e., the system will adjust to changes and learn from them. This may give rise to configurations/topologies which have not yet been conceived.
- *Hybrid information processing:* This means that online information processing can generally be discrete and continuous – depending on the mechanical engineering structure.

23.2
Intelligent Flexible Manufacturing Systems

The market for industrial goods is characterized by ever-shorter product cycles. Because of increasing demands by customers and the rapid pace of development in the area of microelectronics, new products not only have to be developed and marketed faster but also have to be produced more quickly. Apart from speed and efficiency, flexibility therefore is playing an increasingly important role in production processes.

Even some years before the end of the 20th century, the standards achieved in the area of manufacturing and assembly were extremely high as the level of automation continued to increase. New products could be accommodated quickly thanks to CNC-processing machines and powerful robots. And, even now at the turn of the new millennium there is still considerable scope for improvement in the area of in-house material flow and particularly in relation to manufacturing control across machine boundaries. This section looks at different approaches to overcoming existing weaknesses and solving problems in traditional manufacturing systems. The principles presented here are being used increasingly in commercial manufacturing systems and have thus already proven their suitability for practical use.

23.2.1
Paradigm Shifts in Industrial Automation

The central task of industrial automation is to establish a link between the control and monitoring system and the physical process. This generally involves a number of sensors and actuators. Various options are available for connecting sensors and actuators, each of which builds on the next. Figure 23.6 shows a rough split into four generations. First-generation processors emerged on the basis of the simple control and computer technology of the 1960s and 1970s. These processors had a high space requirement and placed stringent demands on the environment. Progressive miniaturization in the 1980s allowed the development and use of programmable logic controllers (PLC), which characterize the second generation of automation. The PLCs represent the first step to decentralized control, firstly because, thanks to their size, they can be integrated into the actual machine and secondly, because of their lower price, they are affordable individually for each machine. The main problems of these two types of processor were the general structure of the point-to-point connections between the systems. The resulting cable trees were expensive and could not be handled.

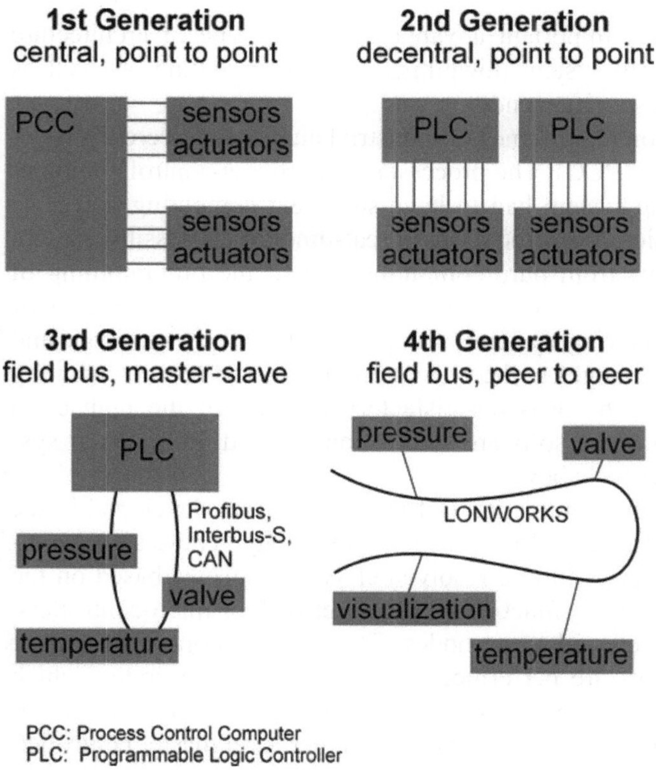

Fig. 23.6 Four generations of communication in the field.

The third generation is characterized by field-bus systems, which began to make their impact on the manufacturing industries in the 1990s. In this context, one of the main considerations in relation to the development of manufacturing systems is cost: The automation effort increases excessively with the number of sensors and actuators. Major factors here include the cost of the circuit materials, the space required for the lines, and above all the installation effort required. Another reason for using field buses is to avoid disruptions caused by electromagnetic radiation, which in particular can corrupt the transmission of analog values. Thanks to the digital transmission of data and the shielding in the field bus, the information is afforded a very high level of protection against such disruptive influences.

The fourth generation of manufacturing automation evolved from the development of information technology at the end of the twentieth century. If mainframe computers still dominated the market for commercial applications in the 1980s, networked PCs and workstations now provide the computing power directly at the user's workplace. In the same way, the automation control changed from a master-slave architecture to a peer-to-peer architecture. The most important advantages of this type of architecture include reduced overall system downtime, ease of upgrading, as well as independence from a specific vendor.

The first-generation mainframes of industrial automation were the process-control computers (PCC). The processor of a process-control computer from a manufacturing system had to have sufficient computing power so that the correct decisions could be made in real-time for all possible system states. However, apart from pure computing power, the programming of the PCC is a more serious issue. Because of the high level of complexity of processes in manufacturing systems, the programs created are large and above all unwieldy. Changes to the system often involved re-implementing the control software. This was unsatisfactory because of the high costs involved in developing the software, and it contradicted the modular system approach as described above.

Decentral Intelligent Automation (DIA) systems were developed in the late 1990s in line with general information technology. These systems allow the problems described to be resolved elegantly. DIA is based on the principle of distributing the functionality to a network of microcontrollers, which are typically referred to as nodes. The computations required for controlling the systems are performed exclusively on these nodes, which assimilate the necessary information via the network. In contrast to classical field-bus networks, there is no need for a higher level master controller, which polls a series of slave controllers. Communication can be organized much more flexibly as a result and can be tailored more closely to the requirements of the process.

Standard PCs are integrated in the network as an interface to the user, although from the perspective of the other nodes these simply represent an additional node (Fig. 23.7).

The advantage of this technology is that a modular set of solution elements can be drawn upon when assembling a system. There is no need to create complex, central programs. Rather, the performance of a manufacturing system is determined more by the assembly of the modules.

23.2.2 Decentral Intelligent Automation

The central idea behind the development of Decentral Intelligent Automation (Gausemeier and Förste 1998) is the intelligent object. Intelligent objects are agents, which respectively model the functions of scheduling, control and monitoring of a single machine or robot in a manufacturing system (Fig. 23.8). Intelligent objects deal much more with the software aspects of control rather than the hardware aspects described above.

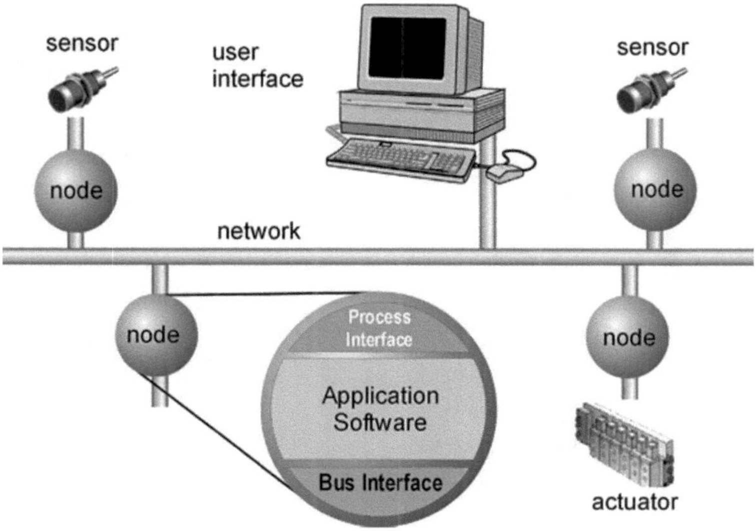

Fig. 23.7 ecentral Intelligent Automation. A system of connected controllers (nodes) which perform a decentral intelligent-automation system.

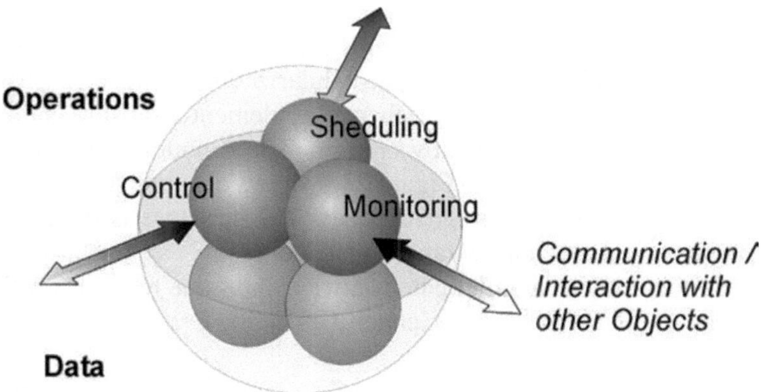

Fig. 23.8 Structure of an intelligent object.

Intelligent objects have a 1:1 relationship with the machines to which they are assigned. The connection is made on the basis of the field bus in the manufacturing network or via an interface from the allocated microcontroller. Beside the interfaces, an intelligent object has, in accordance with the typically object-oriented programming, a data and a function/operation area. The most important task of the operations is to communicate with other objects. The operations are thus divided into three groups in accordance with their tasks (scheduling, control, and monitoring). The data is divided equivalently and assigned to the operations.

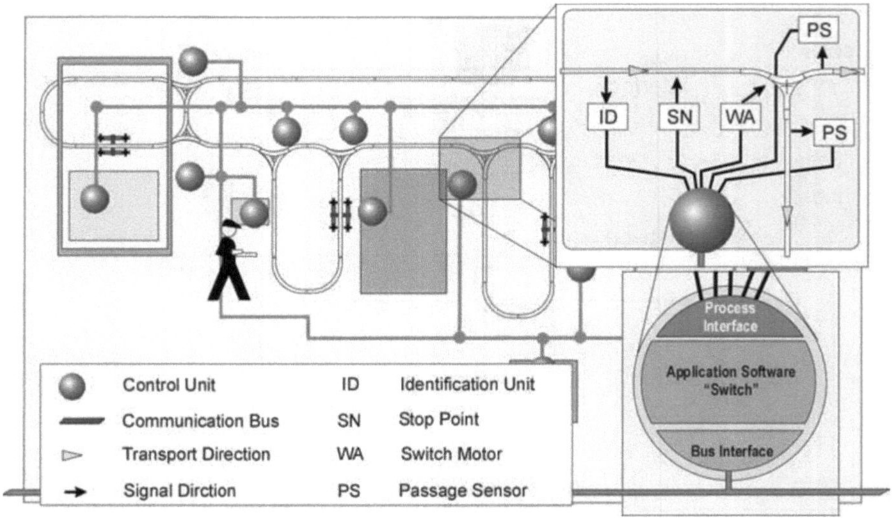

Fig. 23.9 Use of intelligent modules in a Modular Transport System.

The functionality of an intelligent object, which is installed on the node of a DIA field bus, will be described using the example of a rail-borne material-flow system. With a view to high modularization and decentralization, a control node is assigned to each active element (transfer gate, stopping point, etc.) in the material-flow system. These nodes communicate with one another via a field bus. Figure 23.9 shows a sample layout of a material-flow system with the assigned nodes. The structure is demonstrated using the example of a single transfer-gate node.

The principal operation of an inlet gate is always the same regardless of the installed location (Gausemeier and Gehnen 1997a, b):

- Detecting the incoming shuttle,
- Establishing the required transport direction,
- Switching the correct travel way, and
- Enabling passage.

The intelligent object has a process interface and bus interface at its disposal to allow it to fulfill these tasks. The control and monitoring operations use the bus interface to access the connected sensors and actuators, thus allowing tasks one, three, and four to be resolved. The intelligent object uses the bus interface to contact other objects via the communication bus. Depending on the implementation, this also allows the second task to be fulfilled.

The depicted three-layer structure (process interface, application software, and bus interface) of the control module is the same for all nodes. The intelligent objects can be taken from a modular system and placed in the middle layer. This allows ready-made standard components to be used for the software as well as for the hardware. This option can be fully utilized by the vendors of materials-handling technology in that, apart from the mechanical components, the appropriately programmed control system can also be supplied. No programming is therefore required in the manufacturing system, rather the system is simply parameterized.

23.2.3 Example: Modular Transport System

The features described above were used for validation purposes in the manufacturing laboratory at the Heinz Nixdorf Institute in a real manufacturing system. The system comprises several CNC cutting machines, assembly robots, and manually controlled work centers. The central element of the system is a rail-borne transport system from Montech AG (Switzerland). This system is characterized particularly by its modular structure and

Fig. 23.10 Modules of a flexible transport system (from upper left to bottom: shuttle, switch, fixing station). The pictures show the reference-transport system and its components at the Heinz Nixdorf Institute in Paderborn. The system is a Montrac system, built by Montech AG, Switzerland.

its simple range of modules, which nevertheless allow all layouts to be created for a material-flow system. The main system components available are the vehicles (or shuttles), transfer gates, crossings, stopping stations, and straight and curved joining pieces (Fig. 23.10).

A decentralized control architecture was developed on the basis of this, which consistently promotes the modularization of systems to the level of the control hardware. Two different field-bus systems and controls were used in this context. The first implementation saw the LON (Local Operating Network) being used as the field-bus network. This network was developed especially for distributed control systems, which are based on the neuron chip from Echelon. In the second implementation, the same principles of Decentral Intelligent Automation were applied in a network of commercially available PLCs.

23.2.4 Material Flow in Flexible Manufacturing Systems

The architectural design of a control system based on intelligent objects uses the object-oriented approach in software development for the specific area of process control techniques beyond the limits of creating a control system (Gerdes 1997). The concept of object orientation in the context of manufacturing is the basis for a network of communicating intelligent objects embedded in a control system, or the ION (Intelligent Object Network) architecture as it is also referred to (Fig. 23.11).

The most important components of the ION system architecture are naturally the intelligent objects, the agents, and function modules as well as the system interfaces. In addition, communication must be considered from the point of view of the overall system. The ION architecture (Fig. 23.11) comprises a communications platform in which intelligent objects, agents, and function modules are embedded. External interfaces are based on the communications platform and exist for each external communication partner in the control system. The diagram shows an example with three interface modules. The MRP interface (manufacturing resource planning) connects the MRP system with the manufacturing control center. The man-machine interface (MMI) represents the user interface to the control system on the basis of a graphics terminal. The field-bus interface connects the processing machines (symbolized in accordance with the VDI standard), industrial data-capture (IDC) terminals, transport and warehouse-control systems – shown here by way of example via a bus-based communication link – to the control system.

The fundamental components of the ION architecture are the intelligent objects shown as shaded circles. Each intelligent object is connected to a

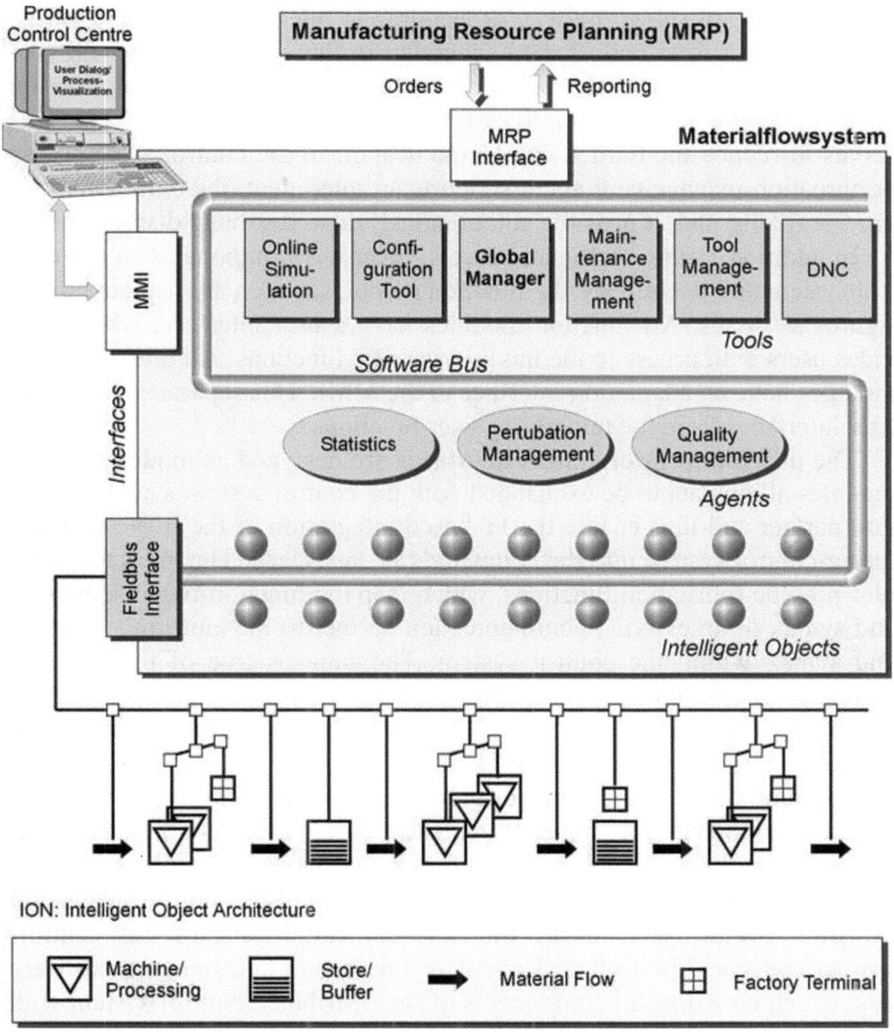

Fig. 23.11 The Intelligent Object Network architecture. A concept for the material-flow control in flexible automated manufacturing systems.

resource in a physically unique way. Apart from the different processing machines, these systems can also be forging manipulators, transport systems, warehouse control systems, or measuring machines. An intelligent object autonomously executes all scheduling, control, and monitoring functions for the resource assigned to it.

Agents are symbolized as ovals in the diagram and are optional in the concept presented here. Agents provide special services, generally for a group of intelligent objects, but also for individual intelligent objects.

Three agents are shown by way of example in the figures. The task of the statistics agent is to collect data generated in the control system and prepare it statistically. The task of the resource agent is to acquire information about the availability of resources for a group of intelligent objects. This serves to reduce the burden of communication in the control system. The perturbation-management agent supports an intelligent object in determining, localizing-and, if possible autonomously, also resolving disruptions.

In addition to the intelligent objects, other key components of the control system architecture are the function modules, which are depicted in the figures as boxes. All function modules have a user interface, which provides users with access to the module-specific functions. All these function modules have an adaptation interface to the MMI. This separates individual user interfaces from the underlying user functions.

The different control system interfaces are designed as modules. These modules allow data to be exchanged with the control system's communication partner and thus enable the IT-based integration of the ION manufacturing-control system into the company's IT landscape. The interface modules provide translation functions, which map the function-based semantics and syntax of an external communication partner to the uniform semantics and syntax within the control system. The same is true of the opposite communication direction. A separate interface module is therefore required to connect each third-party system with a different set of semantics and syntax.

Communication is the basis for the coordinated cooperation between the system components presented. It takes the form of exchanging shared information on the one hand and synchronizing independent processes on the other hand. Communication between intelligent objects and the agents and function modules is therefore the link and cornerstone of this control-system concept. The technical basis for this communication is a software bus, which networks all components of the distributed control system with one another.

23.2.5
Development of Manufacturing Control and Monitoring Systems

When developing intelligent objects for manufacturing control, the developer not only must be familiar with the object's tasks, but also with its environment. He thus needs clear specification of the IT-based manufacturing-control structure. Information technology offers a number of specification methods for this purpose, which are also used to an extent in the engineering disciplines.

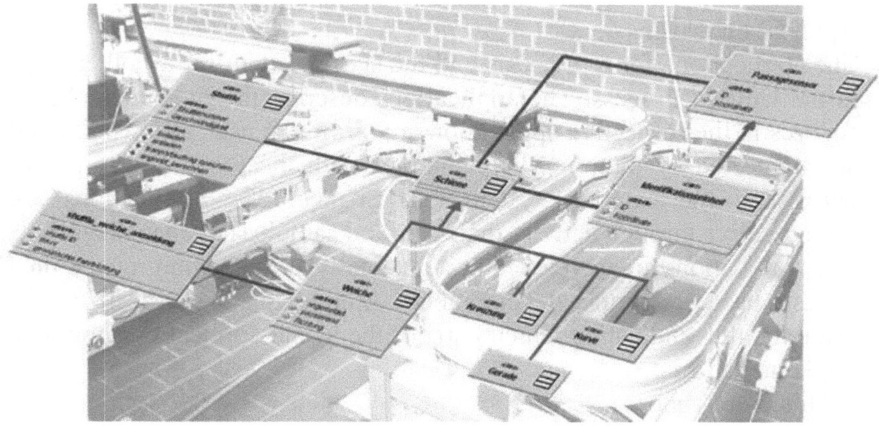

Fig. 23.12 Specification of an intelligent transport system with the Unified Modeling Language (UML).

The Unified Modeling Language (UML) (Booch et al. 1998) is particularly well suited to describing the structure. As a quasi-standard for modeling object-oriented software, it is supported by a wide range of tools. Figure 23.12 shows the model of the material-flow system described above in the UML object-model description. The individual objects in the system, as well as the data, are represented as classes (Gausemeier et al. 2000a,b).

The UML is supported by the Unified Software Development Process (Jacobsen et al. 1999), which describes a modern, universal procedure approach to executing software projects. If the UML is used for the system specification, this process can also be used for modeling the non-software-related components.

The Unified Software Development Process (or Unified Process for short) is the description of a possible approach to developing object-oriented systems. The Unified Process is (Oesterreich 1999):

- Application driven,
- Architecture-centric,
- Iterative, and
- Incremental.

This procedure model is characterized by the division of development activities into smaller units as well as the coordinated interleaving of activities at varying levels of detail. The transition from outline to detail does not occur synchronously for all areas of the system, rather in a differentiated way depending on the priority and the respective problem.

- Application driven
  The applications form the basis and the starting point for all other steps in the Unified Process. The applications describe the fundamental procedures in the respective application area from the user's viewpoint and are supported by plain-text explanations with the focus on the procedures and the external system performance. The applications are brought alive through activity diagrams and sample sequence diagrams.
  The applications are developed systematically at the beginning and then act as a basis for further developments. Applications are prioritized, for example, thus indicating which parts of the actual application should be implemented first.
- Architecture-centric
  The application development model must be able to take into account the features and special characteristics of an existing application architecture. The application architecture defines which artifacts are to be developed. Application architectures are frequently layer models and contain the following layers, for example:
  - Transaction control (information and signal processing)
  - Dialogs, dialog control (user interface with user guidance)
  - Technical classes (mapping of data)
  - Persistence (storing of data, e.g., database)

  In addition, systems are frequently divided into components and subsystems.
  The architecture describes how the entire system is divided into its parts and how these interact with one another. From this it is possible to derive which parts should actually be developed.
  The primary consideration therefore in the development of decentrally controlled manufacturing systems is the choice and programming of the controllers being used and their respective variants. The particularly restricted resources (CPU, memory, etc.) of these controllers must be considered.
- Iterative and incremental
  The system is developed in several successive phases. In the past, the application development followed the so-called waterfall model, i.e., the complete system requirements were established first, then the design was developed for the entire system and finally all parts were implemented.
  The application-driven, iterative/incremental approach differs here fundamentally. First of all, as described above, all application scenarios are elaborated and hence give rise to the outline-requirements description. Large systems are then typically divided into subsystems (modulariza-

tion, system partitioning). These subsystems can then be developed by separate teams.

In order to follow the iterative and incremental model, the interim results of the component teams must be synchronized regularly. While the components should actually be developed largely independently of one another, they should be coordinated centrally so that practical and, above all, planned results can be achieved at defined milestones. These interim results are then validated and verified in internal and external reviews, within the project through test and review preparations as well as quality assurance, and externally by specialist departments, future users, etc.

Each team therefore tries to achieve certain agreed-upon results at defined milestones for all components, with the development process of each individual team also being divided into iterations. Each iteration in turn comprises a detailed planning (defined iteration goals) and an analysis-design-implementation-test sequence (Fig. 23.13).

The Unified Process is a series of processes from the initial definition of requirements to the testing of the finished system. These processes are run in succession until the result complies with the customer's requirements. Iterative means in this context therefore the separation of the development into several steps of the same type. Each iteration creates an interim result. Incremental means that the entire functionality of the system being developed grows with each step.

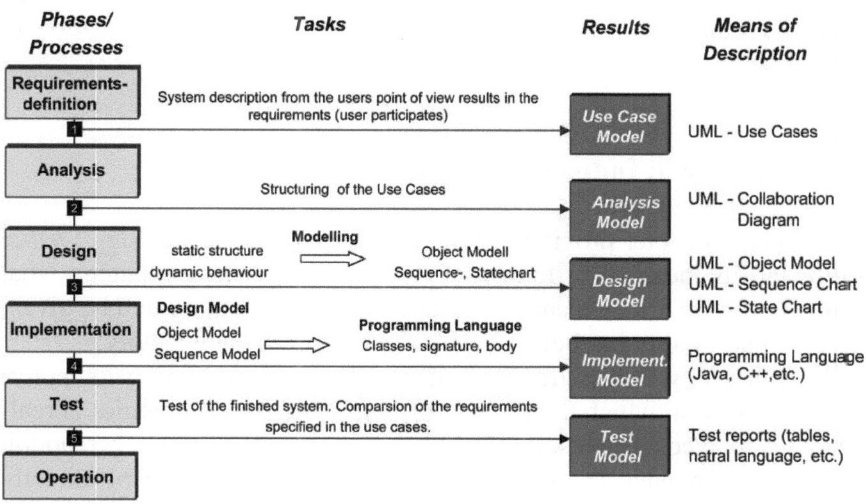

Fig. 23.13 The Unified Process is a flow of processes. Beginning with the requirements definition and resulting in the operation.

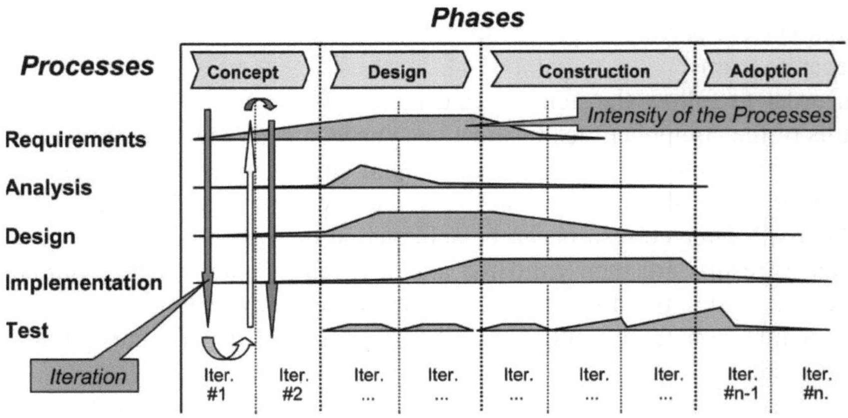

Fig. 23.14 An overview of the Unified Process (Jacobsen et al. 1999).

An iterative/incremental approach therefore implies, in particular, the procedure outlined in Fig. 23.14. In this case, the entire analysis-design-implementation-test sequence from Fig. 23.13 is performed in each iteration (corresponding to a single phase). However, the individual subprocesses are weighted differently. In the early phases (e.g., conceptualization or design), the emphasis is on the definition of requirements. However, decisions may already have to be made at this stage regarding the design, or different implementations may have to be performed (e.g., in order to check the general feasibility of an analysis result). The construction and transition phases primarily involve the implementation. However, the definition of requirements and the analysis are still used here in relation to issues of detail.

23.3 Virtual Reality in Industrial Applications

The following chapter provides a brief overview of the technology *Virtual Reality* and its meaning in Industrial Applications. After explaining what Virtual Reality is from a technical point of view, an example will be given, where Virtual Reality has been used to create a virtual bike factory.

The technology *Virtual Reality* (VR) is 3-D visualization of a computer-generated world in which the entirety of objects of all technologies is realistically displayed. The user is enabled to gain an accurate perception (presentation) of what an object would look like in real life by directly manipulating it in the virtual world (interaction) and being personally integrated (immersion) into it (Ebbesmeyer 97).

The visualization part of VR refers to the computer generating visual, auditory, or other sensual outputs to the user of a world within the computer. This world may be a CAD model, a scientific simulation, or a view into a database. The user can interact with this world and directly manipulate objects within the world. Some worlds are animated by other processes, perhaps physical simulations, or simple animation scripts. Interaction with the virtual world, at least with near real-time control of the viewpoint is essential for a *virtual reality*.

The applications being developed for VR run a wide spectrum, from games to architectural and business planning. Many core applications are virtual representations of our own world that are very similar to our own, like CAD or architectural modeling. Some applications provide ways of viewing from an advantageous perspective not possible with the real world, like scientific, telepresence systems, or air-traffic control systems. Other applications are much different from anything we have ever directly experienced before. These latter applications may be the hardest, and most interesting systems. Examples may be the visualization of the ebb and flow of the world's financial markets or navigating a large corporate-information base.

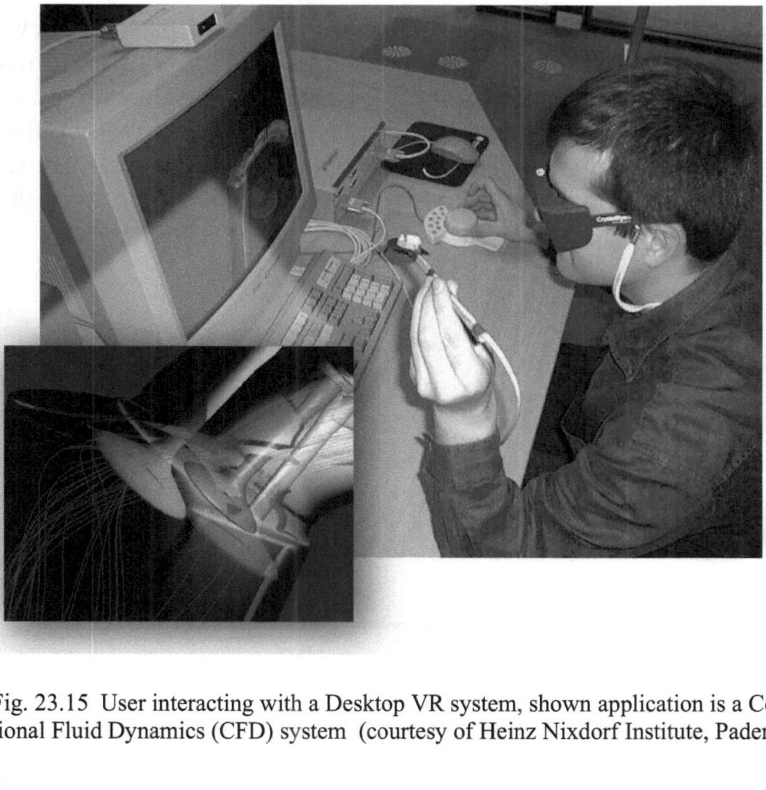

Fig. 23.15 User interacting with a Desktop VR system, shown application is a Computational Fluid Dynamics (CFD) system (courtesy of Heinz Nixdorf Institute, Paderborn).

23.3.1
Types of VR Systems

A major distinction of VR systems is the mode with which they interface to the user. At first we want to explain different types of VR systems and their man-machine interfaces.

Some systems use a conventional computer monitor to display the visual world (Fig. 23.15). This is sometimes called “Desktop VR” or a “Window on a World” (WoW). This concept traces its lineage back through the entire history of computer graphics. In 1965, Ivan Sutherland laid out a research program for computer graphics in a paper called "The Ultimate Display" that has driven the field for the past nearly 30 years (Sutherland 1965).

One must look at a display screen, as a window through which one beholds a virtual world. The challenge to computer graphics is to make the picture in the window look real, sound real and the objects act real.

23.3.1.2
Immersive Systems

The ultimate VR system completely immerses the user's personal viewpoint inside the virtual world. These "immersive" VR systems are often equipped with a Head Mounted Display (HMD). This is a helmet or a face mask that holds the visual and auditory displays. The helmet may be free ranging, tethered, or it might be attached to some sort of a boom armature. HMDs have made improvements in size, quality, and pricing over the years. Figure 23.16 shows two common HMDs from 1994 and 1998.

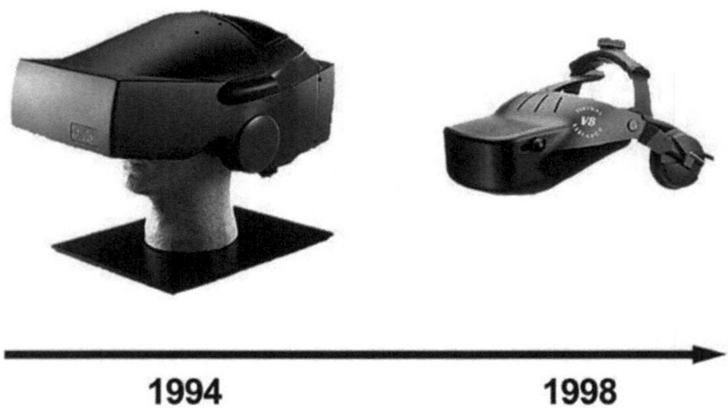

Fig. 23.16 Technical Progress in Head Mounted Displays (courtesy of Virtual Research Systems).

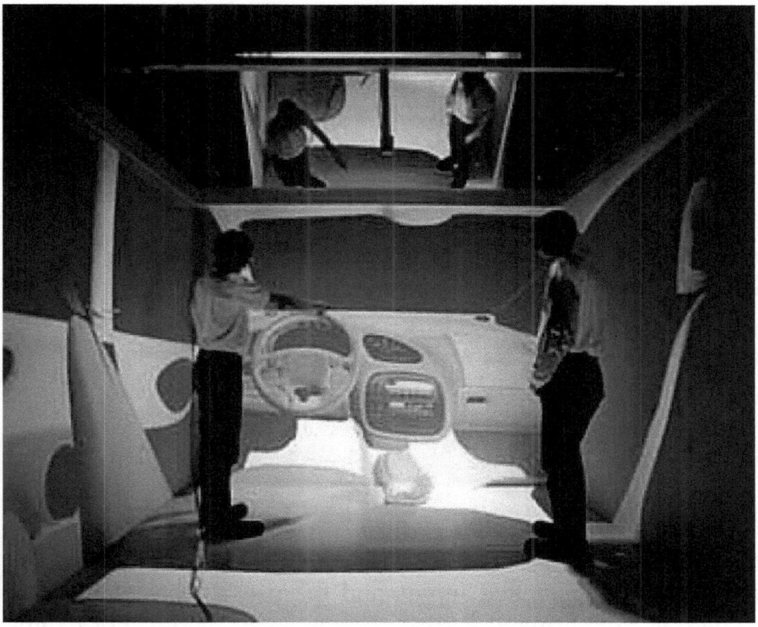

Fig. 23.17 Designer discussing a car interior (courtesy of TAN projection technology).

Most HMDs use two displays and can provide stereoscopic imaging. Others use a single larger display to provide higher resolution, but without the stereoscopic vision.

Most lower cost HMDs, ranging from $3,000 to $10,000, use LCD displays, while others use small CRTs, such as those found in camcorders. The more expensive HMDs use special CRTs mounted alongside the head or optical fibers to pipe the images from non-head-mounted displays. ($60,000 and up). An HMD requires a position tracker in addition to the helmet. Alternatively, the display can be mounted on an armature for support and tracking (a Boom display).

A nice variation of the immersive systems uses multiple large-projection displays to create a 'Cave' or room in which the viewer(s) stand (Fig. 23.17). An early implementation was called "The Closet Cathedral" for the ability to create the impression of an immense environment within a small physical space. The Holodeck used in the television series "Star Trek: The Next Generation" is a far-term extrapolation of this technology.

23.3.1.3 Telepresence

Telepresence is a variation on visualizing complete computer-generated worlds. This technology links remote sensors in the real world with the senses of a human operator. The remote sensors might be located on a robot. They are transmitting audio-visual information collected by the sensors to the user. Robots equipped with telepresence systems have already changed the way deep sea and volcanic exploration is done. NASA is using telerobotics for space exploration. There is currently a joint US/Russian project researching telepresence for space-rover exploration.

23.3.1.4 Mixed Reality / Augmented Reality

While Virtual Reality always uses completely computer-generated environments, introducing the real world can be somewhat helpful. This technology, where computer-generated inputs and the user's view of the real world are merged, is called mixed reality or augmented reality. For example, a surgeon's view of a brain surgery is overlaid with images from earlier CAT (Computerized Axial Tomography) scans and real-time ultrasound images. This helps him to get a better understanding of the case for operation planning.

Most of the augmented-reality systems are mobile, enabling the user to work in non-local environments like industrial plants or outdoor environ

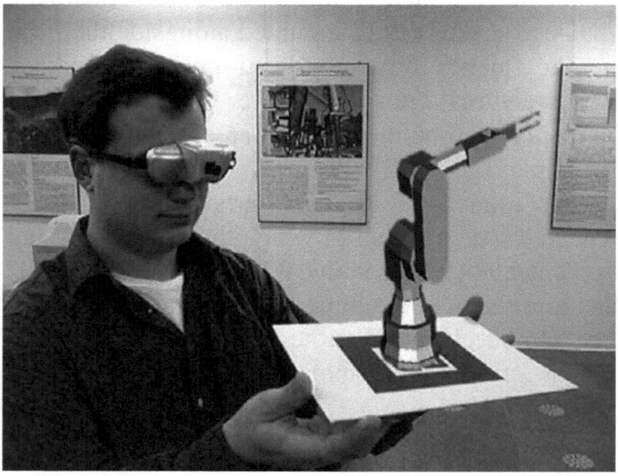

Fig. 23.18 User wearing a mobile Augmented Reality system is interacting with a virtual robot model.

ments. In Fig. 23.18, a user with a mobile system (consisting of a see-through display and a mobile computer) is shown.

23.3.2 VR Hardware

23.3.2.1 Image Generators

One of the most time-consuming tasks in a VR system is the generation of the images. Fast computer graphics opens a very large range of applications aside from VR, so there has been a market demand for hardware acceleration for a long while. Silicon Graphics Inc. (SGI) has made a very profitable business of producing graphics workstations. SGI boxes are some of the most common processors found in VR laboratories and high-end systems. SGI boxes range in price from under $10,000 to over $500,000. The simulator market has produced several companies that build special-purpose computers designed expressly for real-time image generation. These computers often cost several hundreds of thousands of dollars.

Driven by the gaming industry, there are currently a number of vendors selling image-generator cards for PC-level machines. These cards range in price from about $100 up to $1,000. The rendering performance is excellent compared to the pricing. What these image generators do lack are special routines for improving the image quality like anti-aliasing, where different color shades are used to smooth jagged edges of curved and diagonal lines.

23.3.2.2 Manipulation and Control Devices

One key element, for interaction with a virtual world, is a means of tracking the position of a real-world object, such as a head or hand. There are numerous methods for position-tracking and control. Ideally a technology should provide 3 measures for position (X, Y, Z) and 3 measures of orientation (roll, pitch, yaw). One of the biggest problems for position tracking is latency, or the time required to make the measurements and preprocess them before input to the simulation engine.

The simplest control hardware is a conventional mouse, trackball, or joystick. While these are two-dimensional devices, creative programming can use them for 6-D controls. There are a number of 3- and 6-dimensional mice/trackball/joystick devices being introduced to the market at this time. These add some extra buttons and wheels that are used to control not just the XY translation of a cursor, but its Z dimension and rotations in all three directions.

One common VR device is the instrumented glove. The use of a glove to manipulate objects in a computer is covered by a basic patent in the USA. Such a glove is outfitted with sensors on the fingers as well as an overall position/orientation tracker. There are a number of different types of sensors that can be used. VPL Research (holders of the patent) made several DataGloves, mostly using fiber-optic sensors for finger bends and magnetic trackers for overall position.

The drawback of using these devices like gloves in Virtual Reality applications is that the user feels no force while pushing against virtual objects and so has no feedback to his actions. For example, feeling the texture of an object is only possible by using force-feedback devices. There are many force-feedback devices ranging from tactile finger-enclosing gloves to full-body suits.

The CyberGrasp is a lightweight, unencumbering force-reflecting exoskeleton that adds resistive force feedback to each finger (Fig. 23.19). With the CyberGrasp force-feedback system, users are able to explore the physical properties of computer-generated 3-D objects they manipulate in a simulated 'virtual world.'

The concept of an instrumented glove has been extended to other body parts. Full-body suits with position and bend sensors have been used for

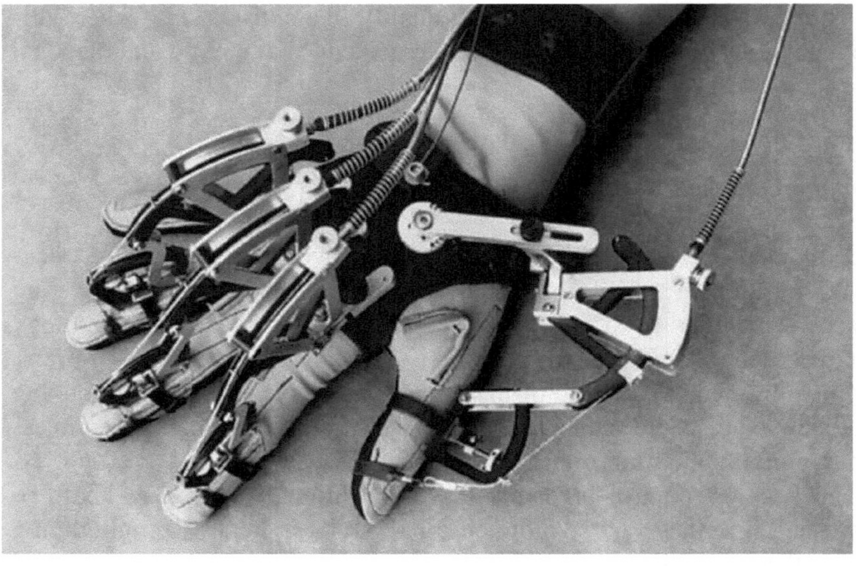

Fig. 23.19 Force-feedback device fitting over a dataglove (courtesy of Virtual Technologies, Inc.).

capturing motion for character animation, control of music synthesizers, etc., in addition to VR applications.

Mechanical armatures can be used to provide fast and very accurate tracking. Such armatures may look like a desk lamp (for basic position/orientation) or they may be highly complex exoskeletons (for more detailed positions). The drawbacks of mechanical sensors are the encumbrance of the device and its restrictions on motion. Exos Systems builds one such exoskeleton for hand control. It also provides force feedback. Shooting Star system makes a low-cost armature system for head tracking. Fake Space Labs and LEEP Systems make much more expensive and elaborate armature systems for use with their display systems.

Ultrasonic sensors can be used to track position and orientation. A set of emitters and receivers are used with a known relationship between the emitters and between the receivers. The emitters are pulsed in sequence and the time lag to each receiver is measured. Triangulation gives the position. Drawbacks to ultrasonics are low resolution, long lag times, and interference from echoes and other noises in the environment. Logitech and Transition State are two companies that provide ultrasonic tracking systems.

Magnetic trackers use sets of coils that are pulsed to produce magnetic fields. The magnetic sensors determine the strength and angles of the fields. Limitations of these trackers are a high latency for the measurement and processing, range limitations, and interference from ferrous materials within the fields. However, magnetic trackers seem to be one of the preferred methods. The two primary companies selling magnetic trackers are Polhemus and Ascension.

Optical position-tracking systems have been developed. One method uses a ceiling grid LED and a head-mounted camera. The LEDs are pulsed in sequence and the cameras image is processed to detect the flashes. Two problems with this method are limited space (grid size) and lack of full motion (rotations). Another optical method uses a number of video cameras to capture simultaneous images that are correlated by high-speed computers to track objects. Processing time (and cost of fast computers) is a major limiting factor here. One company selling an optical tracker is Origin Instruments.

Inertial trackers have been developed that are small and accurate enough for VR use. However, these devices generally only provide rotational measurements. They are also not accurate for slow position changes.

23.3.2.3
Stereo Vision

Stereo vision is often included in a VR system. This is accomplished by creating two different images of the world, one for each eye. The images are computed with the viewpoints offset by the equivalent distance between the eyes. There are a large number of technologies for presenting these two images. The images can be placed side-by-side and the viewer asked (or assisted) to cross their eyes. The images can be projected through differently polarized filters, with corresponding filters placed in front of the eyes. Anaglyph images use red/blue glasses to provide a crude (no color) stereovision.

The two images can be displayed sequentially on a conventional monitor or projection display. Liquid Crystal shutter glasses are then used to shut off alternate eyes in synchronization with the display. When the brain receives the images in rapid enough succession, it fuses the images into a single scene and perceives depth. A fairly high display-swapping rate (min. 60Hz) is required to avoid perceived flicker.

23.3.3
Example: Cyberbikes – A Virtual Enterprise

A modern industrial enterprise is a very complex system. Its objective is to develop, to manufacture, and to deliver affordable industrial products just in time and of high quality. The current marketplace requires flexibility and a high rate of productivity. To meet these challenges business managers have to reengineer and optimize their business processes. The enabling factors in achieving these objectives are the extensive use of information technology (IT) and sophisticated production technology.

Because of its complexity, it is often difficult to understand how a modern industrial enterprise really works. Specifically, how do the various different business processes interact with each other? How do they contribute to the results? How can information technology and smart manufacturing systems contribute to increase productivity? A thorough understanding of how an industrial enterprise and its specific parts work is crucial to those who operate it. It is even more important for those who will operate such an enterprise in the near future – students of mechanical engineering for example.

One obvious solution to the foregoing question is to build one just for the purpose of study. In Germany, for example, a series of model factories have been set up in recent years (Walze 1993). These "CIM-Centers" consist of real systems such as manufacturing systems for the production of parts, assembly systems, and automated-transport systems. The major drawback of building model factories for research, educational, and training purposes is the high cost of the setup, operation, and maintenance of these facilities.

Simulators have successfully been used to teach students about complex systems and to train them to operate these systems. Flight simulators are well-known examples for these types of training devices. Today Virtual Reality can be regarded as an enabling technology very useful in building simulators for a wide variety of special purposes. This section describes a system called CYBERBIKES which immerses a person in a virtual environment which simulates an industrial enterprise (Fig. 23.20). The main purpose of the system is to teach students of mechanical and industrial engineering how a complex industrial enterprise works.

23.3.3.1 Objectives of CYBERBIKES

The basic idea of CYBERBIKES is to supply students with a system which allows them to explore a modern industrial enterprise. An appropriate system should give the user the impression of "really being there." The clear-

Fig. 23.20 A user immersed in the CYBERBIKES virtual environment

est solution to this requirement is to design the core of the system as a Virtual Environment based on realistic-looking 3-D models representing all objects a typical industrial enterprise consists of – i.e., the plant building; the various departments of the production plant with its machine tools, robots, and transportation systems; the offices of engineering, administration, and management; and perhaps even a cafeteria.

Although a realistic-looking 3-D model is the most important part of such a virtual environment, there are other issues which have to be addressed. A simulation of an industrial enterprise must also represent the data to be processed within the production. For example, machine tools and robots have to operate like they do in reality. Thus, there is a need for special simulation models. In addition, the system should also provide 3-D audio to enhance realism.

The challenge is to implement CYBERBIKES as a virtual environment. Every subsystem of CYBERBIKES has to operate in real-time: Graphics rendering of complex 3-D models, acoustic rendering of 3-D sound, accurate simulation of machine tools and robots, clear simulation of material flow, and visualization of the flow of information. Furthermore, there is a need to integrate software which enables a person to control the various business processes. The most realistic solution would be to use real IT applications – i.e., software which is used in the actual world of an industrial enterprise – e.g., manufacturing-control systems (MCS). These systems must rely on complex simulation models which are based on real-world experiences.

23.3.3.2
General Approach

CYBERBIKES is a virtual environment representing a medium-sized enterprise which produces high-quality bicycles. The bike was chosen because it is a well-known product to everybody, though the idea behind the system is not limited to bicycles.

The basic idea of CYBERBIKES is to produce bicycles on demand, i.e., a user can exactly configure the bike he wants and have this configuration manufactured. Because it is very hard to build complex configuration systems in a virtual environment, an Internet-based configuration system has been developed, so that the user starts with a web browser configuring his favorite bike.

For ease of use, the user will be guided during the configuration process. The user can choose from different standard configurations (Fig. 23.21) and stepwise refine them by changing e.g., constructional parameters, or upgrade/downgrade of the configuration (Fig. 23.22).

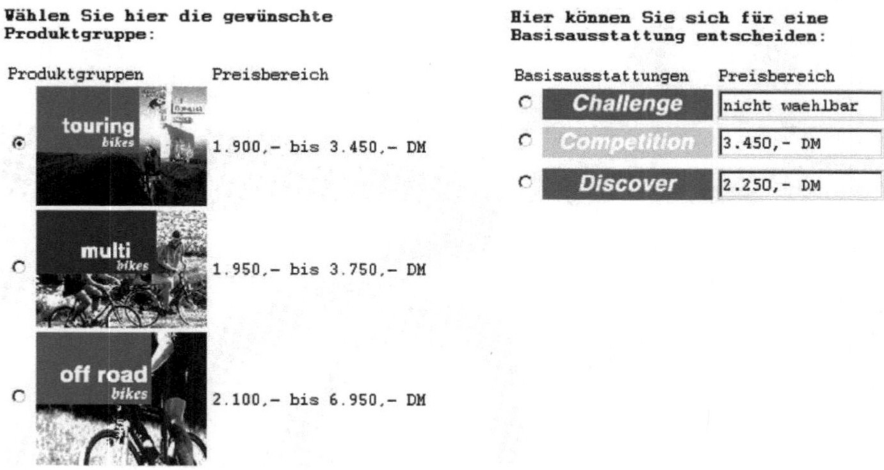

Fig. 23.21 Selecting a bike category by pricing and use criteria.

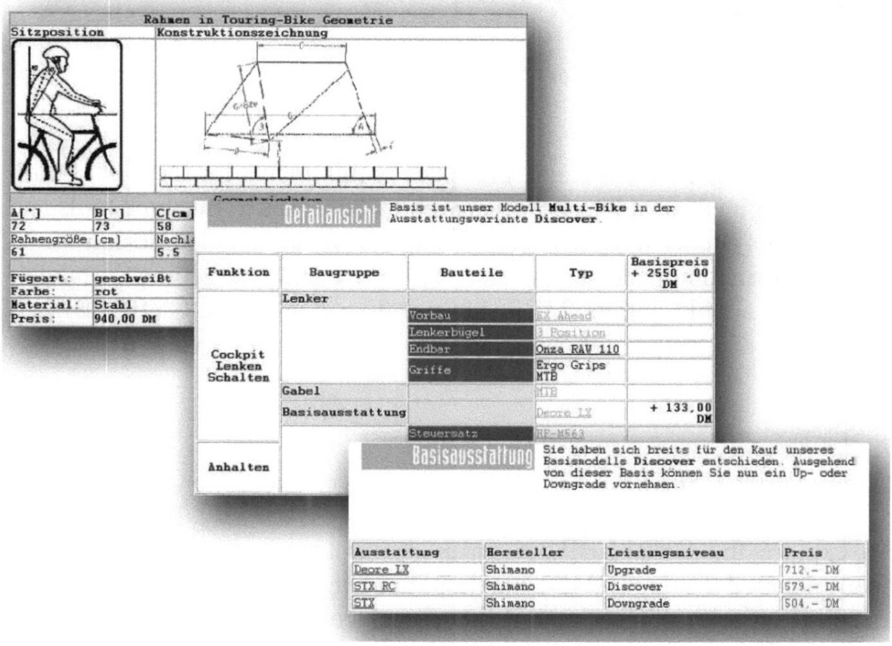

Fig. 23.22 Stepwise refinement of the bike configuration.

CYBERBIKES was designed from scratch, starting with the basic design of the production processes. These are based on actual production processes which can be found in the modern bike-manufacturing industry. The basic design has been reworked in several different forms resulting in

Fig. 23.23 Inner and outer views of CYBERBIKES.

a detailed design. At this stage, the space requirements for the plant building were defined as well as how many and what types of machine tools and production systems were needed. At the end of the detailed design stage, all the information necessary to start geometric and non-geometric modeling were available. Figure 23.23 shows a floor plan of the CYBERBIKES plant from detailed design stage as well as a bird‘s-eye view of the 3-D model.

23.3.3.3
Operating the CYBERBIKES system environment

To explore the CYBERBIKES virtual environment, the user puts on a head-mounted display (HMD) enabling him/her to move through the model and interact with objects within the virtual environment, using a special kind of 3-D input device called a "3-D-mouse." During the walkthrough of the scene, standard magnetic-tracking devices track the user‘s head and hand. The virtual environment itself is generated and controlled by a four-processor graphics workstation (SGI Onyx). The workstation is equipped with two separate graphics subsystems to generate stereoscopic images. Acoustic rendering is carried out by an entry-level workstation (SGI Indy), which drives a 4-speaker ambisonic audio subsystem (Bamford 1995). The

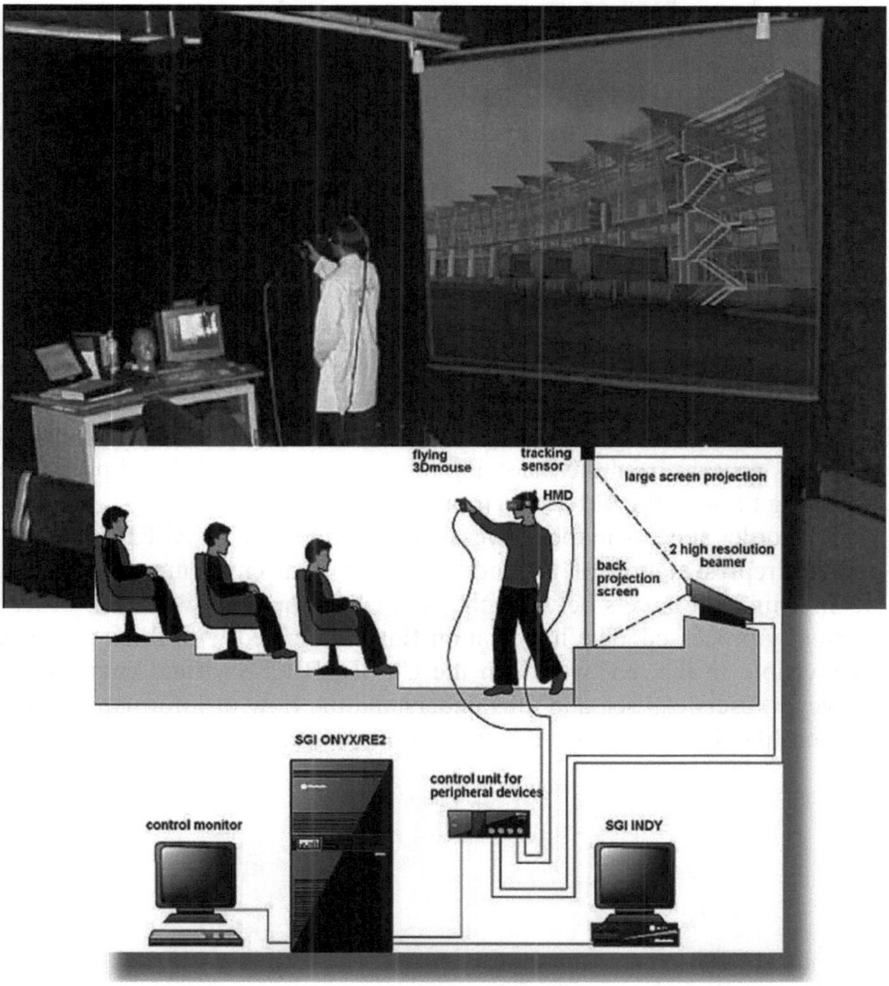

Fig. 23.24 The “Software Theater” used for presentation of the CYBERBIKES virtual environment.

audio host (SGI Indy) and the graphics workstation are connected by a standard Ethernet network.

The system is installed in the "Software Theater," a cinema-like auditorium equipped with 30 seats and a large-scale stereoscopic projection screen (Fig. 23.24). This facility, located in the Heinz Nixdorf Museums Forum next to our institute, was designed by our group for the presentation of CYBERBIKES and other virtual-reality applications.

In the Software Theater, trained instructors offer visitors guided tours to the CYBERBIKES virtual environment. The audience follows the instructor, who is immersed in CYBERBIKES, by simultaneously tracing his/her view presented through the HMD he wears on a large-scale projection screen. Interested users who wish to explore CYBERBIKES themselves can use the HMD while the instructor follows their walkthrough on the projection screen. Thus, the instructor can lead the user to points of interest.

23.3.4
A Sample Bike-Manufacturing Procedure

After the user has configured his favorite bike with the web-based product configurator, an order is generated and sent to the CYBERBIKES system. The simulation system of CYBERBIKES verifies the correctness of the order and acknowledges it. Now the user is encouraged to follow the whole business process from the incoming of the order to the finished product by traveling through the virtual environment.

The order appears in the canvassing department of CYBERBIKES.

The representation of the order in the virtual environment is a small icon. This icon travels between the computers showing where the order is currently processed. The information flow associated with these processes is invisible in the real world. In the CYBERBIKES virtual environment, however, users can see and thus understand the flow of information. Information-like data is visualized by means of metaphors (Fig. 23.25).

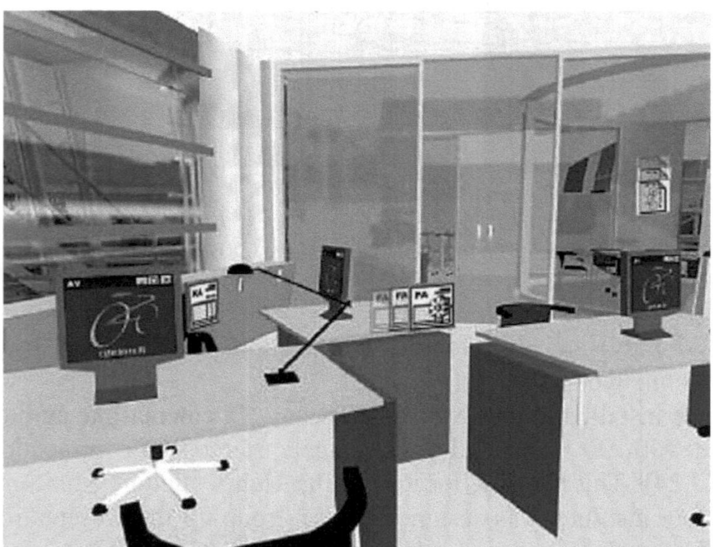

Fig. 23.25 Visualization of information flow within CYBERBIKES (Orders moving from host to host).

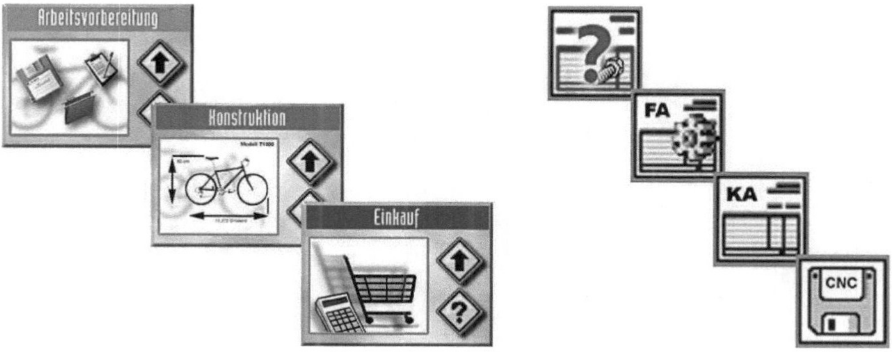

Fig. 23.26 Entering the virtual environment of CYBERBIKES.

Fig. 23.27 Pedal-crank manufacturing system.

The visitor of CYBERBIKES is now able to watch the processing of customer orders. These orders generate manufacturing instructions which are subsequently forwarded to manufacturing units consisting of machine tools and industrial robots. They can be watched interactively in the virtual environment as follows:

The user equipped with HMD and flying mouse enters the virtual main entrance of CYBERBIKES and goes towards the production area (Fig. 23.26).

Fig. 23.28 Gear-rim manufacturing system.

The pedal-crank and gear-rim production consists of numerical-controlled machine tools, industrial robots, and transportation devices. Using a manufacturing-control system (MCS), the user is able to start a production process and watch it while it works. The results of the interaction are visualized by the MCS immediately. The user is also able to modify the production process currently running. He can stop machine tools and change production tasks. His actions are controlled by the MCS. The MCS is an example of software that can be found in a real-world enterprise. We have integrated such a system into CYBERBIKES (Gausemeier et al 1996). The graphical-user interface of the MCS is displayed on a virtual monitor (Figs. 23.27 and 23.28).

The user immersed in CYBERBIKES interacts with the system by means of a touch-screen metaphor using his hand which is tracked by a tracking device. His real hand is represented in the virtual environment by a 3-D hand icon. Due to the low resolution of both the HMD and the tracking system, this virtual touch-screen allows only simple interactions, which are sufficient for most users. If all features of the MCS are needed, it has to be run on an extra screen. The machine tools in the virtual environment are then controlled from a real terminal.

23.3.5
System Design and Implementation

23.3.5.1
Geometric Modeling

All geometric modeling was done with 3-D CAD tools. The plant building itself and the interior of the offices were designed by the department of architecture at the University of the Arts in Berlin, Germany. All other models were created at our institute. The machine tools and industrial robots were modeled with the help of drawings and photographs of real machine tools used for bicycle production. The CAP-tool RobCad was used for modeling and simulation of the manufacturing systems. The geometry data was then converted to a real-time data format using several converters. The simulation of the production process has been recorded to generate animation data for the virtual machine tools and robots.

Realism of the models was enhanced by applying realistic materials and texture maps onto the surfaces of the models. Furthermore, a lot of time was spent optimizing the models for use in a real-time environment. The whole geometric data set of CYBERBIKES consists of 1,357,000 polygons. 129 megabytes are required to store it in a compact binary-data format (108 MB geometry files and 21 MB texture files). Figure 23.29 show a selection of different inner and outer views from CYBERBIKES. They underline the complexity and the level of realism of the geometric models used in the CYBERBIKES virtual environment.

Fig. 23.29 Different views of the CYBERBIKES virtual environment.

Fig. 23.30 Hierarchical Structure of Production Control in Industrial Enterprises.

Real industrial enterprises are operated by actual human labor. Thus, human beings should also be included in a realistic simulation of an industrial enterprise. Realistic models of human beings for the use in real-time environments are difficult to create and rather resource-consuming (Durlach and Mavor 1995). This is especially true when numerous representations are needed simultaneously. The simulation of "real" human beings for VR is still a very challenging task. Therefore, in the majority of existing VR-applications humans look and behave like toys. We decided to make light of this situation by specifically modelling humans as toys. We believe this is an acceptable solution for our application.

23.3.5.2
Non-Geometric Modeling

Data-processing in modern industrial enterprises is hierarchically structured into four management layers as shown in Fig. 23.30, ranging from issuing production orders with declared delivery dates to controlling particular machine components, and where each layer processes specific tasks.

- Planning and Disposition Layer: The management of primary data and material management occurs at this layer. Customer orders are managed and handed on to the next layer.
- Manufacturing-Control Layer: At this layer, the customer orders, received from the planning and disposition layer, are transformed into manufacturing orders. At this stage, planning of details is carried out – i.e., staffing schedules of particular machines are generated by considering the predetermined delivery dates taken from customer orders. As a part of the manufacturing layer, NC-programs are passed to machines on demand. In addition, the manufacturing layer ensures the efficiency of the production run. Reports of disturbances are collected and forwarded to subsequent applications.
- Process-Control Layer: This layer is responsible for the coordination of the production equipment and the specific transport systems. Records of material flow in the production process are kept here, and processing data of the individual machines is registered for quality control.
- Machine-Control Layer: This is the lowest management layer. It contains the controls for all automated manufacturing machines.

23.3.5.3
Real-Time Environment

The simulation model for the CYBERBIKES virtual environment comprises of all geometric and non-geometric models as shown in Fig. 23.31.

The simulation model is executed through the real-time environment. The most important requirement of the system is to deliver real-time frame rates (more than 15 frames/sec). This is crucial to maintain the illusion of an interactive virtual world. The hardware resources needed to process the complete CYBERBIKES data-set in real-time is nearly a magnitude greater than what state-of-the-art graphics hardware delivers today.

In recent years, the increase in complexity of virtual environments has outpaced the improvements of graphics hardware. Thus there is ongoing research in reducing the geometric complexity of a given scene for real-time rendering. Among others, the most important strategies are level-of-

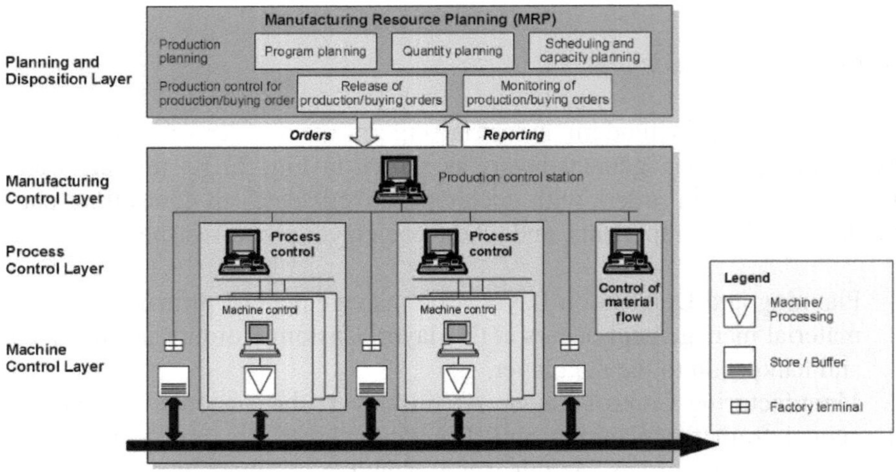

Fig. 23.31 The parts of the simulation model.

detail hierarchies (Clark 1976; Funkhouser and Sequin 1993), visibility culling (Airey et al. 1990; Teller 1992), occlusion culling (Coorg and Teller 1997; Zhang et al. 1997), and image-based rendering utilizing texture maps (Shade et al. 1996; Ebbesmeyer 1997). The CYBERBIKES virtual environment utilizes several of these advanced-rendering techniques. The implementation is based on a special purpose Application Programming Interface (API) for real-time rendering (Rohlf and Helman 1994). The number of polygons to be drawn in every frame could be reduced to less than 50,000 (of 1.3 million), which allows a smooth walkthrough on a state-of-the-art ONYX 2 IR. The frame rate is still acceptable on the Reality Engine 2 machine, which is installed in the Software Theater.

A first version of CYBERBIKES was completed in October 1996 (Gausemeier et al. 1996). This version consisted of the plant building and 20% of the production systems. At that time, no offices were included. The virtual environment has been continuously updated since then. Several production systems as well as office areas have been added. CYBERBIKES has been completed in December 1997. Furthermore, it was presented at the SGI booth during the CeBit computer show of 1998 as a major attraction.

Since the initial version of the system was available, CYBERBIKES has been presented and tested in the Software Theater mentioned above. Since then more than 50,000 people have seen these presentations. Ten percent of the visitors used the HMD themselves. The majority of those who have immersed themselves in CYBERBIKES got a positive impression. There where a number of complaints about the low resolution of the HMD. Several users had difficulties finding their way around when being immersed in the CYBERBIKES virtual environment. It is obviously necessary to have efficient navigational aids in virtual environments of such complexity. A detailed study on the usability of the system to find out whether it meets its design goals will be done in the near future.

We have described CYBERBIKES, a real-time simulation of a typical industrial enterprise, based on a 3-D-virtual environment. The core of the system is a realistic 3-D model of an industrial plant. In addition, the business processes, like the order-to-invoice process and the manufacturing process, are modeled. The user is able to explore the model of the virtual enterprise by means of immersing himself/herself in this environment using a head-mounted display. Flows of information associated with these processes are visualized. Because the information flow is invisible in reality, the animation of the information flows in the virtual environment helps the users to understand easily how an enterprise works. This is essential for the purpose of education as well as for optimizing the organization by management.

Future work on the system will be focused on adding navigational aids. This could be accomplished by introducing agents and smart maps. A further area of research will include extending CYBERBIKES to a multi-user environment where more than one user can be immersed simultaneously.

Literature

- Airey JM, Rohlf JH, Brooks FP (1990) Towards image realism with interactive update rates in complex virtual building environments. Computer Graphics, volume 24, 1990
- Bamford J (1995) An analysis of ambisonic sound systems of first and second order. University of Waterloo, Waterloo
- Booch G, Rumbaugh J, Jacobsen I (1998) The unified modelling language user guide. Addison-Wesley, Reading, Massachusetts
- Clark JH (1976) Hierarchical geometric models for visible surface algorithms. Communications of the ACM, volume 10: 547–554
- Coorg S, Teller S (1997) Real-time occlusion culling for models with large occluders. In: Proceedings ACM Symposium on Interactive 3D Graphics '97, pp 83–90
- Durlach NI, Mavor AS (eds) (1995) Virtual Reality: Scientific and Technological Challenges. National Research Council, National Academy Press, Washington DC
- Ebbesmeyer P (1997) Dynamische Texturwände - Ein Verfahren zur echtzeitorientierten Bildgenerierung für Virtuelle Umgebungen technischer Objekte. HNI-Verlagsschriftenreihe, Paderborn
- Funkhouser TA, Sequin CH (1993) Adaptive display algorithm for interactive frame rates during visualization of complex virtual environments. Proceedings of SIGGRAPH '93 volume 27:247–254
- Gausemeier J, von Bohuszewicz O, Gehnen G, Grafe M (1996) Cyberbikes: an immersive virtual environment controlled by real CIM-applications. In: Proceedings of the FIVE 1996 Conference, Pisa

- Gausemeier J, Förste S (1998) Dezentral automatisiertes Materialflußsystem. ZwF '98, pp 411–414
- Gausemeier J, Gehnen G (1997a) Intelligent material flow by decentral control networks. Proceedings of The Second World Congress on Intelligent Manufacturing Processes & Systems, Budapest, Hungary, June 10–13, Springer, Berlin Heidelberg New York
- Gausemeier J, Gehnen G (1997b) Integrated network for decentral intelligent manufacturing control and automation. Preprints of 4th IFAC Workshop on Intelligent Manufacturing Systems IMS 97, Seoul, Korea, July 21–22, IFAC
- Gausemeier J, Kespohl H D, Molt T (2000a) Specification of a decentral intelligent manufacturing control system with the Unified Modelling Language (UML). ICCIM, Singapore
- Gausemeier J, Kespohl H, Molt T (2000b) Softwaretechnik im Maschinenbau. ZwF Jahrg. 95 (6): 282–288
- Gerdes KH (1997) Concept of an architecture for a flexible manufacturing control system (German: Architekturkonzeption für Fertigungsleitsysteme der flexible automatisierten Fertigung). HNI-Verlagsschriftenreihe, Paderborn
- Jacobsen I, Booch G, Rumbaugh J (1999) The unified software development process. Addison-Wesley, Reading, Massachusetts
- Lückel J (1992) The concept of mechatronics function modules applied to compound active suspension systems. International Symposium for Vehicle Dynamics, Herbertor, CSFR
- Lückel J, Henke M, Liu X, Grotstollen H, Jäker KP (2000) Design of a railway carriage driven by a linear motor with active suspension/tilt module. 9th IFAC Symposium on Control in Transportation Systems 2000, Braunschweig
- Naumann R (2000) Modellierung und Verarbeitung vernetzter intelligenter mechatronischer Systeme. HNI-Verlagsschriftenreihe, Paderborn
- Oesterreich B (1999) Erfolgreich mit Objektorientierung. Oldenbourg Wissenschaftsverlag, München
- Pahl G, Beitz W (1997) Konstruktionslehre – Methoden und Anwendung. 4. Aufl., Springer Verlag
- Rohlf J, Helman J (1994) IRIS Performer: A high performance multiprocessing toolkit for real-time 3D graphics. In: Proceedings of SIGGRAPH '94, Orlando, Florida, pp 381–395
- Shade J, Lishinski D, Salesin D H, DeRose T, Snyder J (1996) Hierarchical image caching for accelerated walkthroughs of complex environments. In: Proceedings of SIGGRAPH '96, New Orleans, Louisiana, pp 75–82
- Sutherland I (1965) The ultimate display. In: Proceedings of the International Federation of Information Processing Congress, New York City, New York, volume 2:506–508
- Teller SJ (1992) Visibility computations in densely occluded polyhedral environments. University of California, Berkeley
- Walze, H (ed) (1993) Computer Integrated Manufacturing (CIM) – Results from projects within the manufacturing technologies program 1988–1992. PFT-Report. Kernforschungszentrum Karlsruhe GmbH
- Zhang H, Manocha D, Hudso T, Hoff KE (1997) Visibility culling using hierarchical occlusion maps. In: Proceedings of SIGGRAPH '97, Los Angeles, California, pp 77–88

MIX
Papier aus verantwortungsvollen Quellen
Paper from responsible sources
FSC® C105338

If you have any concerns about our products,
you can contact us on
ProductSafety@springernature.com

In case Publisher is established outside the EU,
the EU authorized representative is:
Springer Nature Customer Service Center GmbH
Europaplatz 3, 69115 Heidelberg, Germany

Printed by Libri Plureos GmbH
in Hamburg, Germany